扶松柏 编著

图像识别技术与实战

(OpenCV+dlib+Keras+Sklearn+TensorFlow)

清华大学出版社
北京

内 容 简 介

本书循序渐进地讲解了使用 Python 语言实现图像视觉识别的核心知识，并通过具体实例的实现过程演练了图像视觉识别的方法和流程。全书共 12 章，分别讲解了图像识别技术基础，scikit-image 数字图像处理，OpenCV 图像视觉处理，dlib 机器学习和图像处理算法，face_recognition 人脸识别，Scikit-Learn 机器学习和人脸识别，TensorFlow 机器学习和图像识别，国内常用的第三方人脸识别平台，AI 人脸识别签到打卡系统(PyQt5+百度 AI+OpenCV-Python+SQLite3 实现)，基于深度学习的 AI 人脸识别系统(Flask+OpenCV-Python+Keras+Sklearn 实现)，AI 考勤管理系统(face-recognition+Matplotlib+Django+Scikit-Learn+dlib 实现)，AI 小区停车计费管理系统。全书讲解简洁而不失技术深度，内容丰富全面，并且易于阅读，以极简的文字介绍了复杂的案例，是学习 Python 图像视觉识别的实用教程。

本书适用于已经了解 Python 语言基础语法的读者，并且适应于希望进一步提高自己 Python 开发水平的读者，还可以作为大专院校相关专业的师生用书和培训机构的专业教材。

图书在版编目(CIP)数据

图像识别技术与实战：OpenCV+dlib+Keras+Sklearn+TensorFlow/扶松柏编著. 一北京：清华大学出版社，2021.12

ISBN 978-7-302-59408-6

Ⅰ. ①图…　Ⅱ. ①扶…　Ⅲ. ①图像识别　Ⅳ. ①TP391.41

中国版本图书馆 CIP 数据核字(2021)第 212828 号

责任编辑： 魏　莹　刘秀青
装帧设计： 李　坤
责任校对： 周剑云
责任印制： 杨　艳
出版发行： 清华大学出版社
　　网　　址： http://www.tup.com.cn, http://www.wqbook.com
　　地　　址： 北京清华大学学研大厦 A 座　　**邮　　编：** 100084
　　社 总 机： 010-62770175　　**邮　　购：** 010-62786544
　　投稿与读者服务： 010-62776969, c-service@tup.tsinghua.edu.cn
　　质量反馈： 010-62772015, zhiliang@tup.tsinghua.edu.cn
印 装 者： 三河市金元印装有限公司
经　　销： 全国新华书店
开　　本： 185mm×230mm　　**印　　张：** 20.75　　**字　　数：** 453 千字
版　　次： 2022 年 1 月第 1 版　　**印　　次：** 2022 年 1 月第 1 次印刷
定　　价： 89.00 元

产品编号：092611-01

前言

图像识别是人工智能的一个重要领域，是指利用计算机对图像进行处理、分析和理解，以识别各种不同模式的目标和对象的技术，并能对质量不佳的图像进行一系列的增强与重建技术手段，从而有效改善图像质量。

随着计算机及信息技术的迅速发展，图像识别技术的应用逐渐扩展到多个领域，尤其是在面部及指纹识别、卫星云图识别及临床医疗诊断等方面都日益发挥着重要作用。此外在日常生活中，图像识别技术的应用也十分普遍，比如车牌捕捉、商品条码识别及手写识别等。随着图像识别技术的逐渐发展并不断完善，未来它将具有更加广泛的应用前景。

本书的特色

1．内容全面

本书详细讲解 Python 图像视觉识别所需要的开发技术，循序渐进地讲解了这些技术的使用方法和技巧，帮助读者快速步入 Python 数据分析的高手之列。

2．实例驱动教学

本书采用理论加实例的教学方式，通过实例实现了知识点的横向切入和纵向比较，让读者有更多的实践演练机会，并且可以用不同的方式展现一个知识点的用法，真正实现了拔高的教学效果。

3．详细介绍图像视觉识别的流程

本书从一开始便对图像视觉识别的流程进行了详细介绍，而且在讲解中结合了多个实用性很强的数据分析项目案例，带领读者掌握 Python 图像视觉识别的相关知识，以解决实际工作中的问题。

4. 扫描二维码，获取配书学习资源

本书正文中每个二级标题后都放了一个二维码，读者可通过该二维码在线观看视频讲解，帮助读者深入理解书中的案例，提升学习效率。此外，读者还可以扫描下方的二维码获取书中案例源代码。

扫码下载全书源代码

5. 贴心提示和注意事项提醒

本书根据需要在各章安排了很多“注意”“说明”和“技巧”等小板块，让读者可以在学习过程中更轻松地理解相关知识点及概念，更快地掌握个别技术的应用技巧。

本书内容

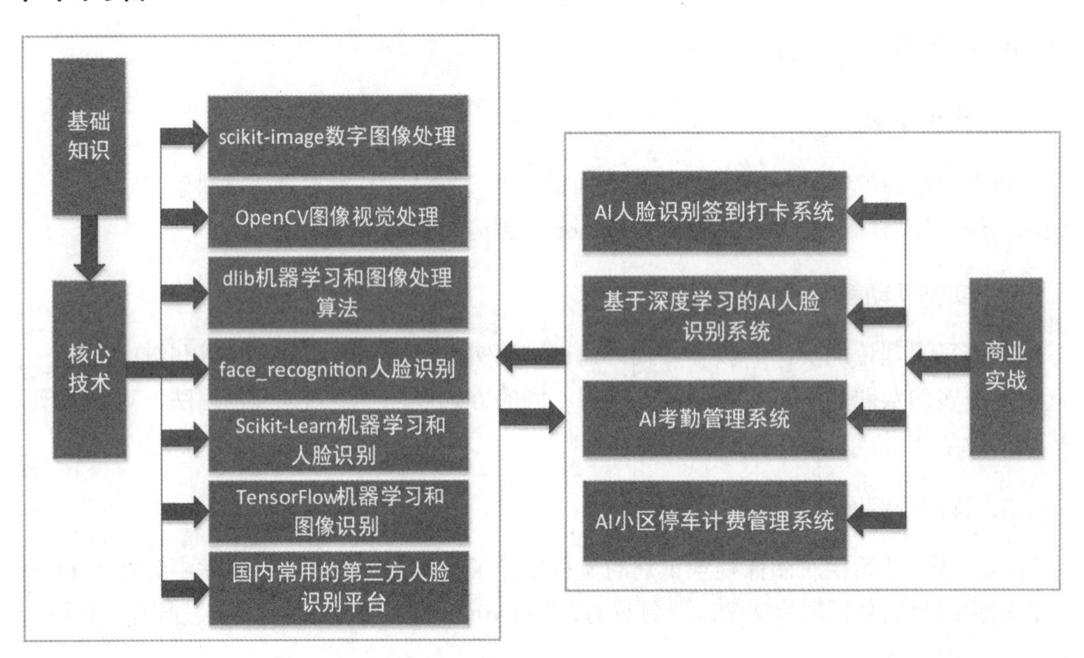

本书读者对象

- 软件工程师。
- Python 语言初学者。
- 专业图像视觉识别人员。
- 数据库工程师和管理员。
- 研发工程师。
- 大学及中学教育工作者。

致谢

本书在编写过程中，得到了清华大学出版社各位编辑的大力支持，正是各位专业人士的求实、耐心和效率，才使得本书能够在这么短的时间内出版。另外也十分感谢我的家人给予的巨大支持。本人水平毕竟有限，书中存在纰漏之处在所难免，真诚感谢读者提出的宝贵意见或建议，以便修订并使之更臻完善。

最后感谢您购买本书，希望本书能成为您编程路上的领航者，祝您阅读快乐！

编　者

Contents 目录

第1章

图像识别技术基础

图像识别是指利用计算机对图像进行处理、分析和理解，以识别各种不同模式的目标和对象的技术，是应用深度学习算法的一种实践应用。在本章的内容中，将详细讲解图像识别技术的基础知识，为读者步入本书后面知识的学习打下基础。

1.1 图像识别概述

扫码观看本节视频讲解

当我们看到一个东西时，大脑会迅速判断是不是见过这个东西或者类似的东西。这个过程有点儿像搜索，我们把看到的东西和记忆中相同或相类似的东西进行匹配，从而识别它。用机器进行图像识别的原理也是类似的，是通过分类并提取重要特征而排除多余的信息来识别图像。机器的图像识别和人类的图像识别原理相近，过程也大同小异。只是技术的进步让机器不但能像人类一样认花认草认物认人，还开始拥有超越人类的识别能力。

1.1.1 什么是图像识别

图像识别是人工智能的一个重要领域，是指利用计算机对图像进行处理、分析和理解，以识别各种不同模式的目标和对象的技术，并对质量不佳的图像执行一系列的增强与重建技术手段，从而有效改善图像质量。

本书所讲解的图像识别并不是用人类肉眼识别，而是借助计算机技术进行识别。虽然人类的识别能力很强大，但是对于高速发展的社会，人类自身识别能力已经满足不了我们的需求，于是就产生了基于计算机的图像识别技术。这就像人类研究生物细胞，完全靠肉眼观察细胞是不现实的，这样自然就产生了显微镜等用于精确观测的仪器。通常一个领域出现固有技术无法解决的需求时，就会产生相应的新技术。图像识别技术也是如此，此技术的产生就是为了让计算机代替人类去处理大量的物理信息，解决人类无法识别或者识别率特别低的信息。

1.1.2 图像识别的应用

移动互联网、智能手机以及社交网络的发展带来了海量图片信息，不受地域和语言限制的图片逐渐取代了烦琐而微妙的文字，成为传词达意的主要媒介。但伴随着图片成为互联网中的主要信息载体，很多难题也随之出现。当信息由文字记载时，我们可以通过关键词搜索轻松找到所需内容并进行任意编辑。但是当信息是由图片记载时，我们无法对图片中的内容进行检索，从而影响了从图片中找到关键内容的效率。图片给我们带来了快捷的信息记录和分享方式，却降低了信息检索效率。在这个环境下，计算机的图像识别技术就显得尤为重要。

(1) 图像识别的初级应用。

在现实应用中，图像识别的初级应用主要是娱乐化、工具化，在这个阶段用户主要借助图像识别技术来满足某些娱乐化需求。例如，百度魔图的“大咖配”功能可以帮助用户找到与其长相最匹配的明星，百度的图片搜索可以找到相似的图片；Facebook 研发了根据相片进行人脸匹配的 DeepFace；国内专注于图像识别的创业公司旷视科技成立了 VisionHacker 游戏工作室，借助图形识别技术研发移动端的体感游戏。

在图像识别的初级应用中还有一个非常重要的细分领域——OCR(Optical Character Recognition，光学字符识别)，是指光学设备检查纸上打印的字符，通过检测暗、亮的模式确定其形状，然后用字符识别方法将形状翻译成计算机文字的过程，就是计算机对文字的阅读。借助 OCR 技术，可以将文字和信息提取出来。在这方面，国内产品包括百度的涂书笔记和百度翻译等；而谷歌借助经过 DistBelief 训练的大型分布式神经网络，对于 Google 街景图库的上千万门牌号的识别率超过 90%，每天可识别百万门牌号。

(2) 图像识别的高级应用。

图像识别的高级应用主要是指成为拥有视觉的机器，当机器真正具有了视觉之后，它们完全有可能代替我们去完成这些行动。目前的图像识别应用就像是盲人的导盲犬，在盲人行动时为其指引方向；而未来的图像识别技术将会同其他人工智能技术融合在一起成为盲人的全职管家，不需要盲人进行任何行动，而是由这个管家帮助其完成所有事情。

举个例子，如果图像识别是一个工具，就如同我们在驾驶汽车时佩戴谷歌眼镜，它将外部信息进行分析后传递给我们，我们再依据这些信息做出行驶决策；而如果将图像识别利用在机器视觉和人工智能上，这就如同谷歌的无人驾驶汽车，机器不仅可以对外部信息进行获取和分析，还全权负责所有的行驶活动，让我们得到完全解放。

1.2 图像识别原理

图像识别的发展经历了 3 个阶段，分别是文字识别、数字图像处理与识别和物体识别，具体说明如下。

扫码观看本节视频讲解

- 文字识别的研究是从 1950 年开始的，一般是识别字母、数字和符号，从印刷文字识别到手写文字识别，应用非常广泛。
- 数字图像处理与识别的研究开始于 1965 年。数字图像与模拟图像相比具有易存储、传输快速、可压缩、传输过程中不易失真、处理方便等巨大优势，这些都为图像识别技术的发展提供了强大的动力。

- 物体识别主要指的是对三维世界的客体及环境的感知和认识，属于高级的计算机视觉范畴。它是以数字图像处理与识别为基础的结合人工智能、系统学等学科的研究方向，其研究成果被广泛应用在各种工业及探测机器人上。

概括来说，图像识别的过程主要包括如下 4 个步骤。

(1) 获取信息：主要是指将声音和光等信息通过传感器向电信号转换，也就是对识别对象的基本信息进行获取，并将其向计算机可识别的信息转换。

(2) 信息预处理：主要是指采用去噪、变换及平滑等操作对图像进行处理，基于此使图像的重要特点突出。

(3) 抽取及选择特征：主要是指在模式识别中，抽取及选择图像特征，概括而言就是识别图像具有种类多样的特点，如采用一种方式分离，就要识别图像的特征，获取特征也被称为特征抽取；在特征抽取中所得到的特征也许对此次识别并不都是有用的，这个时候就要提取有用的特征，这就是特征的选择。特征抽取和选择在图像识别过程中是非常关键的技术之一，所以对这一步的理解是图像识别的重点。

(4) 设计分类器及分类决策：其中设计分类器就是根据训练对识别规则进行制定，基于此识别规则能够得到特征的主要种类，进而使图像识别的辨识率不断提高，此后再通过识别特殊特征，最终实现对图像的评价和确认。

在使用计算机进行图像识别的应用中，计算机首先就能够完成图像分类并选出重要信息，排除冗余信息，根据这一分类计算机就能够结合自身记忆存储结合相关要求进行图像的识别，这一过程本身与人脑识别图像并不存在着本质差别。对于图像识别技术来说，其本身提取出的图像特征直接关系着图像识别能否取得较为满意的结果。

值得读者注意的是，归根结底，毕竟计算机不同于人类的大脑，所以计算机提取出来的图像特征存在着不稳定性，这种不稳定性往往会影响图像识别的效率与准确性。在这个时候，在图像识别中引入 AI 技术就变得十分重要了。

1.3 图像识别技术

计算机的图像识别技术就是模拟人类的图像识别过程，在图像识别的过程中进行模式识别是必不可少的。在本节的内容中，将详细讲解现实中主流的图像识别技术。

扫码观看本节视频讲解

1.3.1 AI(人工智能)

人工智能就是我们平常所说的AI，全称是Artificial Intelligence。人工智能是研究、开发用于模拟、延伸和扩展人类智能的理论、方法、技术及应用系统的一门新的技术科学。人工智能由不同的领域组成，如机器学习，计算机视觉，等等。总的说来，人工智能研究的一个主要目标是使机器能够胜任一些通常需要人类智能才能完成的复杂工作。

人工智能单从字面上应该理解为人类创造的智能。那么什么是智能呢？如果人类创造了一个机器人，这个机器人能有像人类一样甚至超过人类的推理、学习、感知、处理等这些能力，那么就可以将这个机器人称为是一个有智能的物体，也就是人工智能。

现在通常将人工智能分为弱人工智能和强人工智能，我们看到电影里的一些人工智能大部分都是强人工智能，它们能像人类一样思考如何处理问题，甚至能在一定程度上做出比人类更好的决定，它们能自适应周围的环境，解决一些程序中没有遇到的突发事件，具备这些能力的就是强人工智能。但是在目前的现实世界中，大部分人工智能只是实现了弱人工智能，这能够让机器具备观察和感知的能力，在经过一定的训练后能计算一些人类不能计算的事情，但是它并没有自适应能力，也就是它不会处理突发的情况，只能处理程序中已经写好的、已经预测到的事情，这就叫作弱人工智能。

在AI领域中，图像识别技术占据着极为重要的地位，而随着计算机技术与信息技术的不断发展，AI中的图像识别技术的应用范围不断扩展，例如IBM的Watson医疗诊断、各种指纹识别、支付宝的面部识别，以及百度地图中全景卫星云图识别等都属于这一应用的典型。目前，AI这一技术已经应用于日常生活之中，图像识别技术将来定会有着较为广泛的运用。

1.3.2 机器学习

机器学习(Machine Learning，ML)是一门多领域交叉学科，涉及概率论、统计学、逼近论、凸分析、算法复杂度理论等多门学科。机器学习专门研究计算机怎样模拟或实现人类的学习行为，以获取新的知识或技能，重新组织已有的知识结构使之不断改善自身的性能。

机器学习是一类算法的总称，这些算法企图从大量历史数据中挖掘出其中隐含的规律，并用于预测或者分类，更具体地说，机器学习可以看作是寻找一个函数，输入是样本数据，输出是期望的结果，只是这个函数过于复杂，以至于不太方便形式化表达。需要注意的是，机器学习的目标是使学到的函数很好地适用于“新样本”，而不仅仅是在训练样本上表现很好。学到的函数适用于新样本的能力，称为泛化(Generalization)能力。

机器学习有一个显著的特点，也是机器学习最基本的做法，就是使用一个算法从大量的数据中解析并得到有用的信息，再从中学习，然后对之后真实世界中会发生的事情进行预测或作出判断。机器学习需要海量的数据来进行训练，并从这些数据中得到有用的信息，然后反馈到真实世界的用户中。

我们可以用一个简单的例子来说明机器学习，假设在天猫或京东购物的时候，天猫和京东会向我们推送商品信息，这些推荐的商品往往是我们很感兴趣的，这个过程是通过机器学习完成的。其实这些推送商品是京东和天猫根据我们以前的购物订单和经常浏览的商品记录而得出的结论，可以从中得出商城中的哪些商品是我们感兴趣并且会有可能购买，然后将这些商品定向推送给我们。

1.3.3 深度学习

深度学习(Deep Learning，DL)是机器学习领域中一个新的研究方向，它被引入机器学习使其更接近于最初的目标——人工智能(AI)。深度学习是学习样本数据的内在规律和表示层次，这些学习过程中获得的信息对诸如文字、图像和声音等数据的解释有很大的帮助。它的最终目标是让机器能够像人一样具有分析学习能力，能够识别文字、图像和声音等数据。深度学习是一个复杂的机器学习算法，在语音和图像识别方面取得的效果远远超过先前相关技术。

深度学习在搜索技术、数据挖掘、机器学习、机器翻译、自然语言处理、多媒体学习、语音、推荐和个性化技术，以及其他相关领域都取得了很多成果。深度学习使机器模仿视听和思考等人类的活动，解决了很多复杂的模式识别难题，使得人工智能相关技术取得了很大进步。

1.3.4 基于神经网络的图像识别

神经网络图像识别技术是一种比较新型的图像识别技术，是在传统的图像识别方法和基础上融合神经网络算法的一种图像识别方法。这里的神经网络是指人工神经网络，也就是说这种神经网络并不是动物本身所具有的真正的神经网络，而是人类模仿动物神经网络后人工生成的。在神经网络图像识别技术中，遗传算法与 BP 网络相融合的神经网络图像识别模型是非常经典的，在很多领域都有它的应用。

在图像识别系统中利用神经网络系统，一般会先提取图像的特征，再利用图像所具有的特征映射到神经网络进行图像识别分类。以汽车拍照自动识别技术为例，当汽车通过的时候，检测设备就会启用图像采集装置来获取汽车正反面的图像。获取图像后，必须将图像上传到计算机进行保存以便识别。最后，车牌定位模块就会提取车牌信息，对车牌上的字符进行识别并显示最终的结果。在对车牌上的字符进行识别的过程中，就用到了基于模板匹配算法和基于人工神经网络算法。

1.3.5 基于非线性降维的图像识别

计算机的图像识别技术是一个异常高维的识别技术，不管图像本身的分辨率如何，其产生的数据经常是多维性的，这给计算机的识别带来了非常大的困难。想让计算机具有高效的识别能力，最直接有效的方法就是降维。降维分为线性降维和非线性降维。例如主成分分析(PCA)和线性奇异分析(LDA)等就是常见的线性降维方法，它们的特点是简单、易于理解。但是通过线性降维处理的是整体的数据集合，所求的是整个数据集合的最优低维投影。

经过验证，这种线性降维策略计算复杂度高而且占用相对较多的时间和空间，因此就产生了基于非线性降维的图像识别技术，它是一种极其有效的非线性特征提取方法。此技术可以发现图像的非线性结构，而且可以在不破坏其本征结构的基础上对其进行降维，使计算机的图像识别在尽量低的维度上进行，这样就提高了识别速率。例如，人脸图像识别系统所需的维数通常很高，其复杂度之高对计算机来说无疑是巨大的“灾难”。由于在高维度空间中人脸图像的不均匀分布，使得人类可以通过非线性降维技术来得到分布紧凑的人脸图像，从而提高人脸识别技术的有效性。

第2章

scikit-image 数字图像处理

scikit-image 是一款著名的 Python 第三方库，主要功能是处理数字图像。scikit-image 是基于库 Scipy 实现的，它将图片作为 NumPy 数组进行处理。在本章的内容中，将详细讲解使用 scikit-image 处理图像的知识，为读者步入本书后面知识的学习打下基础。

2.1 scikit-image 基础

scikit-image 是一个图像处理库，是采用 Python 语言编写的。scikit-image 提供了多个处理图像的模块，开发者只需调用这些模块即可。

扫码观看本节视频讲解

2.1.1 安装 scikit-image

要想使用库 scikit-image，需要安装如下所示的库。

- Python 2.6 及以上版本。
- NumPy 1.6.1 及以上版本。
- Cython 0.21 及以上版本。
- Six 1.4 及以上版本。
- SciPy 0.9 及以上版本。
- Matplotlib 1.1.0 及以上版本。
- NetworkX 1.8 及以上版本。
- Pillow 1.7.8 及以上版本。

我们可以使用 pip 命令安装 scikit-image，安装命令如下：

```
pip install scikit-image
```

虽然可以在没有虚拟环境的情况下使用 pip，但还是建议大家创建一个虚拟环境，这样可以在干净的 Python 环境中安装 scikit-image，也可以很容易地删除。

在 scikit-image 的官方网站提供了详细的教程，开发者可以了解使用 scikit-image 的方式，官网网站教程的网址是：

https://scikit-image.org/docs/stable/index.html

2.1.2 scikit-image 中的模块

在库 scikit-image 中提供了很多个模块，每个模块包含很多属性和方法，通过这些属性和方法可以实现图像处理功能。scikit-image 中的常用模块如下所示。

- io：读取、保存和显示图片及视频。

- color：颜色空间变换。
- data：提供一些测试图片和样本数据。
- filters：实现图像增强、边缘检测、排序滤波器、自动阈值等功能。
- draw：操作于 NumPy 数组上的基本图形绘制，包括线条、矩阵、圆和文本等。
- transform：几何变换和其他变换，例如旋转、拉伸和 Radon(拉东)变换等。
- exposure：图像强度调整，例如直方图均衡化等。
- feature：特征检测和提取，例如纹理分析等。
- graph：图论操作，例如最短路径。
- measure：图像属性测量，例如相似度和轮廓。
- morphology：形态学操作，例如开闭运算和骨架提取等。
- novice：简化的用于教学目的的接口。
- restoration：修复算法，例如去卷积算法、去噪等。
- segmentation：图像分割为多个区域。
- util：通用工具。
- viewer：简单图形用户界面，用于可视化结果和探索参数。

在本章后面的内容中，将详细讲解使用 scikit-image 中的模块处理图像的知识。

2.2 显示图像

通过使用 scikit-image 可以方便地显示不同图片的样式，例如灰度图像、颜色通道等。在本节的内容中，将详细讲解使用 scikit-image 显示图像的知识。

扫码观看本节视频讲解

2.2.1 使用 skimage 读入并显示外部图像

通过使用 io 和 data 子模块，可以实现图像的读取、显示和保存功能，其中 io 模块用于图片的输入输出操作。为了便于开发者练习，scikit-image 提供了 data 模块，在里面嵌套了一些素材图片，开发者可以直接使用。

例如在下面的实例文件 skimage01.py 中，演示了使用 skimage 读入外部图像并显示的过程。

源码路径：daima\2\2-1\skimage01.py

```
from skimage import data,io
img = io.imread('111.jpg')
io.imshow(img)
io.show()
```

执行代码后会显示读取的外部图片，如图 2-1 所示。

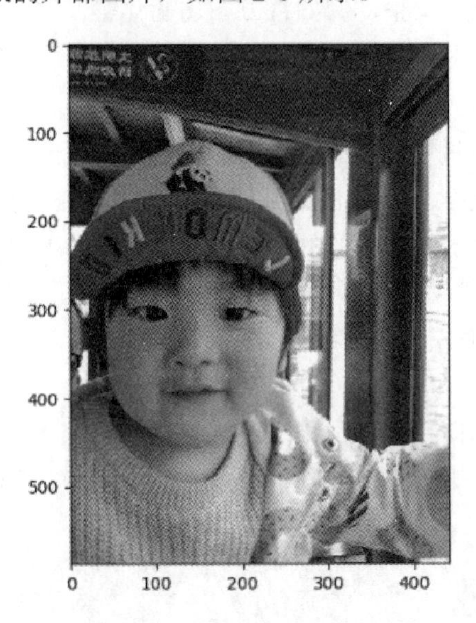

图 2-1　显示读取的外部图片

2.2.2　读取并显示外部灰度图像

如果想读取并显示灰度图效果，可以将函数 imread()中的 as_grey 参数设置为 True，此参数的默认值为 False。例如在下面的实例文件 skimage02.py 中，演示了使用 skimage 读取并显示外部灰度图像的过程。

源码路径：daima\2\2-2\skimage02.py

```
from skimage import data,io
img = io.imread('123.jpg',as_gray=True)
io.imshow(img)
io.show()
```

执行代码后会显示读取的灰度外部图片，如图 2-2 所示。

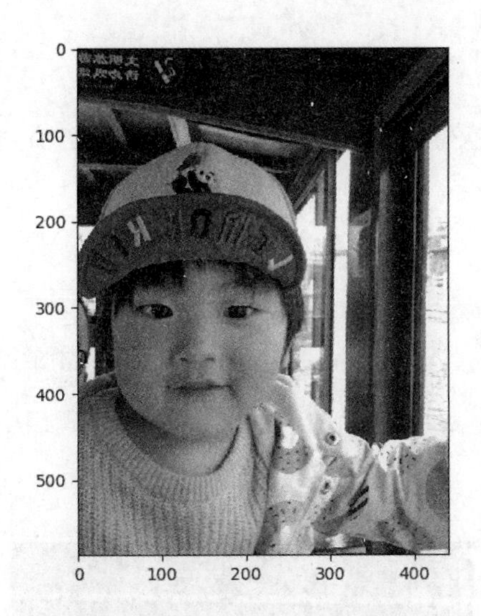

图 2-2　显示外部灰度图片

2.2.3　显示内置星空图片

在 data 子模块中内置了一些素材图片，开发者可以直接使用，具体说明如下所示。

- astronaut：宇航员。
- coffee：一杯咖啡。
- lena：美女。
- camera：拿相机的人。
- coins：硬币。
- moon：月亮。
- checkerboard：棋盘。
- horse：马。
- page：书页。
- chelsea：小猫。
- hubble_deep_field：星空。
- text：文字。
- clock：时钟。
- immunohistochemistry：结肠。

例如在下面的实例文件 skimage03.py 中，演示了使用 skimage 读取并显示内置星空图片的过程。

源码路径：daima\2\2-3\skimage03.py

```
from skimage import io, data
from skimage import data_dir
image = data.hubble_deep_field()                    #读取内置星空图片
io.imshow(image)
io.show()                                           #显示图片
print(data_dir)                                     #打印素材图片的路径
```

在上述代码中，图片名对应的就是函数名。执行代码后会读取并显示星空图片，如图 2-3 所示。

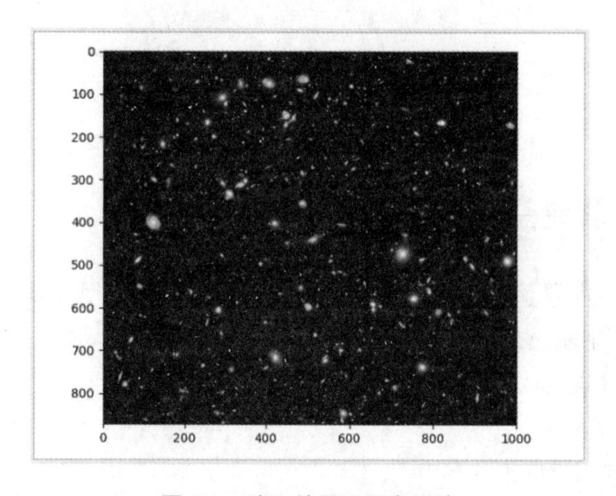

图 2-3　读取并显示星空图片

在 data 子模块中，图片名对应的就是函数名，例如 camera 图片对应的函数名为 camera()。这些素材图片被保存在 skimage 的安装目录下，具体路径名称为 data_dir。通过上述代码的最后一行代码，打印输出了 data_dir 目录的具体路径，例如在笔者机器中执行代码后会输出：

```
C:\Users\apple\AppData\Roaming\Python\Python36\site-packages\skimage\data
```

2.2.4　读取并保存内置星空图片

使用 io 子模块中的函数 imsave(fname,arr)可实现保存图片的功能。其中参数 fname 表示保存的路径和名称，参数 arr 表示需要保存的数组变量。例如在下面的实例文件 skimage04.py 中，演示了使用 skimage 读取并保存内置星空图片的过程。

源码路径：daima\2\2-4\skimage04.py

```
from skimage import io,data
img = data.hubble_deep_field()
io.imshow(img)
io.show()
io.imsave('hubble_deep_field.jpg', img)      #保存图片
```

执行代码后会将读取的星空素材图片保存到本地，名称为 hubble_deep_field.jpg，执行效果如图 2-4 所示。

图 2-4　保存在本地的 hubble_deep_field.jpg

2.2.5 显示内置星空图片的基本信息

通过如下所述的成员可以获取图片的相关信息。

- type()：显示类型。
- shape()：显示尺寸。
- shape()：图片宽度。
- shape()：图片高度。
- shape()：图片通道数。
- size()：显示总像素个数。
- max()：最大像素值。
- min()：最小像素值。
- mean()：像素平均值。

例如在下面的实例文件 skimage05.py 中，演示了使用 skimage 显示内置星空图片基本信

息的过程。

源码路径：daima\2\2-5\skimage05.py

```
from skimage import io,data
img=data.hubble_deep_field()
print(type(img))            #显示类型
print(img.shape)            #显示尺寸
print(img.shape[0])         #图片宽度
print(img.shape[1])         #图片高度
print(img.shape[2])         #图片通道数
print(img.size)             #显示总像素个数
print(img.max())            #最大像素值
print(img.min())            #最小像素值
print(img.mean())           #像素平均值
```

执行代码后会输出：

```
<class 'numpy.ndarray'>
(872, 1000, 3)
872
1000
3
2616000
255
0
19.1544541284
```

2.2.6 实现内置猫图片的红色通道的效果

当将外部或内置素材图片读入程序中后，这些图片是以 NumPy 数组的形式存在的。正因为如此，NumPy 数组的一切相关操作功能对这些图片也是适用的。例如对数组元素的访问，实际上等同于对图片像素的访问。例如在下面的实例文件 skimage06.py 中，演示了使用 skimage 输出内置猫图片的红色通道效果的过程。

源码路径：daima\2\2-6\skimage06.py

```
from skimage import io, data
img = data.chelsea()
#输出图片的R通道中的第20行30列的像素值
pixel = img[20, 30, 1]
print(pixel)
#显示猫图片的红色通道的图片
R = img[:, :, 0]
```

```
io.imshow(R)
io.show()
```

执行代码后会输出内置猫图片的 R 通道中的第 20 行 30 列的像素值为“129”，并显示内置猫图片的红色通道的效果，如图 2-5 所示。

图 2-5 内置猫图片的红色通道的效果

2.3 常见的图像操作

通过前面的学习，读者已经了解了使用 scikit-image 显示图像的知识。其实 scikit-image 的功能远不止如此，在本节的内容中，将进一步讲解 scikit-image 处理图像的知识。

扫码观看本节视频讲解

2.3.1 对内置猫图片进行二值化操作

除了显示图像外，开发者还可以使用 scikit-image 修改图片。例如在下面的实例文件 skimage07.py 中，演示了使用 skimage 对内置猫图片进行二值化操作的过程。

源码路径：daima\2\2-7\skimage07.py

```
from skimage import io, data, color
img=data.chelsea()
img_gray=color.rgb2gray(img)
rows,cols=img_gray.shape
for i in range(rows):
        for j in range(cols):
```

```
            if (img_gray[i,j]<=0.5):
                img_gray[i,j]=0
            else:
                img_gray[i,j]=1
io.imshow(img_gray)
io.show()
```

在上述代码中，使用模块 color 的函数 rgb2gray()将彩色三通道图片转换成灰度图。转换后的结果为 float64 类型的数组，具体范围在[0,1]之间。执行代码后会输出显示二值化后的图片效果，如图 2-6 所示。

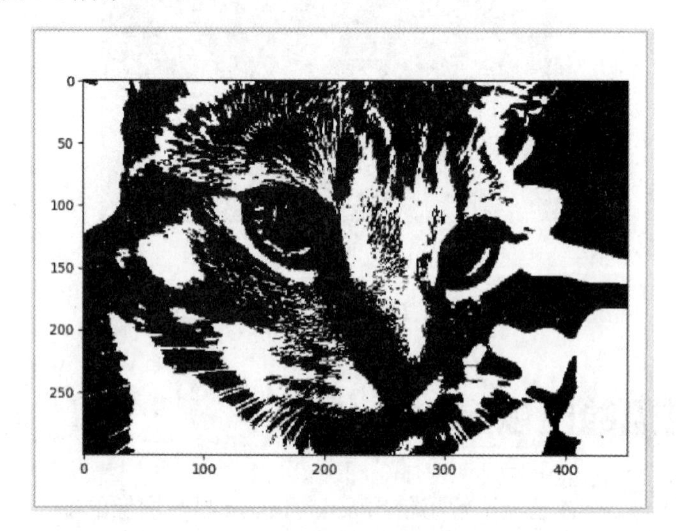

图 2-6　二值化后的效果

2.3.2　对内置猫图片进行裁剪处理

在下面的实例文件 skimage08.py 中，演示了使用 scikit-image 对内置猫图片进行裁剪处理的过程。

源码路径：daima\2\2-8\skimage08.py

```
from skimage import io, data
img = data.chelsea()
roi = img[150:250, 200:300, :]
io.imshow(roi)
io.show()
```

执行代码后会输出显示裁剪之后的效果，如图 2-7 所示。

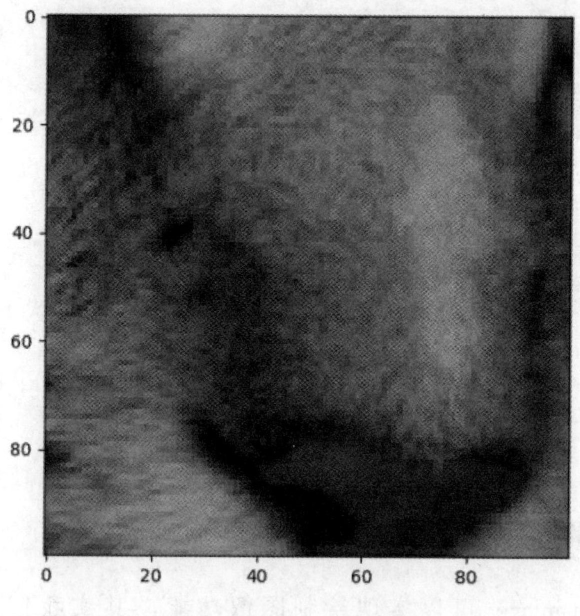

图 2-7　裁剪之后的效果

2.3.3 将 RGB 图转换为灰度图

借助于 scikit-image，可以通过转换颜色空间的方式来实现数据类型转换功能。现实中常用的颜色空间有灰度空间、RGB 空间、HSV 空间和 CMYK 空间，在转换颜色空间以后，所有的数据类型都变成了 float 类型。例如在下面的实例文件 skimage11.py 中，演示了使用 scikit-image 将 RGB 图转换为灰度图的过程。

源码路径：daima\2\2-9\skimage11.py

```
from skimage import io, data, color
image = data.chelsea()
image_grey = color.rgb2gray(image)
io.imshow(image_grey)
io.show()
```

执行代码后会显示将内置猫图片转换成灰度图的效果，如图 2-8 所示。

图 2-8　转换成灰度图后的效果

2.3.4　使用 skimage 实现绘制图片功能

通过使用 scikit-image 可以实现绘制图像功能，其实我们前面多次用到过的 io.imshow(image)函数实现的就是绘图功能。例如在下面的实例文件 skimage12.py 中，演示了使用 skimage 实现绘制图片功能的本质。

源码路径：daima\2\2-10\skimage12.py

```
from skimage import io, data
image = data.chelsea()
axe_image = io.imshow(image)
print(type(axe_image))
io.show()
```

执行代码后会输出绘制图片的功能类是：

```
<class 'matplotlib.image.AxesImage'>
```

matplotlib 是一个专业绘图的库，其相关内容将在本书后面的章节中进行讲解。通过上述实例可知，无论我们利用 skimage.io.imshow()还是 matplotlib.pyplot.imshow()绘制图像，最终都是调用的 matplotlib.pyplot 模块。

2.3.5　使用 subplot()函数绘制多视图窗口

在使用 scikit-image 绘制图片的过程中，我们可以用 matplotlib.pyplot 模块下的 figure()函数来创建显示一个窗口。但是使用 figure()函数创建窗口时存在一个弊端，那就是只能显示一幅图片。如果想要显示多幅图片，则需要将这个窗口再划分为几个子图，在每个子图中显示不同的图片。此时可以使用 subplot()函数来划分子图，此函数的格式为：

```
matplotlib.pyplot.subplot(nrows, ncols, plot_number)
```

参数说明如下。

- nrows：子图的行数。
- ncols：子图的列数。
- plot_number：当前子图的编号。

例如在下面的实例文件 skimage13.py 中，演示了使用 subplot()函数绘制多通道图像的过程。

源码路径：daima\2\2-11\skimage13.py

```
from skimage import data,io
import matplotlib.pyplot as plt
from pylab import mpl
#下面两行代码能保证汉字正确显示
mpl.rcParams['font.sans-serif'] = ['FangSong'] # 指定默认字体
mpl.rcParams['axes.unicode_minus'] = False # 解决保存图像是负号'-'显示为方块的问题
image = io.imread('111.jpg')

plt.figure(num='cat', figsize=(8, 8))  # 创建一个名为cat的窗口，并设置大小

plt.subplot(2, 2, 1)
plt.title('原始图像')
plt.imshow(image)

plt.subplot(2, 2, 2)
plt.title('R通道')
plt.imshow(image[:, :, 0])

plt.subplot(2, 2, 3)
plt.title('G通道')
plt.imshow(image[:, :, 1])

plt.subplot(2, 2, 4)
plt.title('B通道')
plt.imshow(image[:, :, 2])

plt.show()
```

执行代码后不但显示原始图片，而且还会显示另外 3 个通道的子视图，如图 2-9 所示。

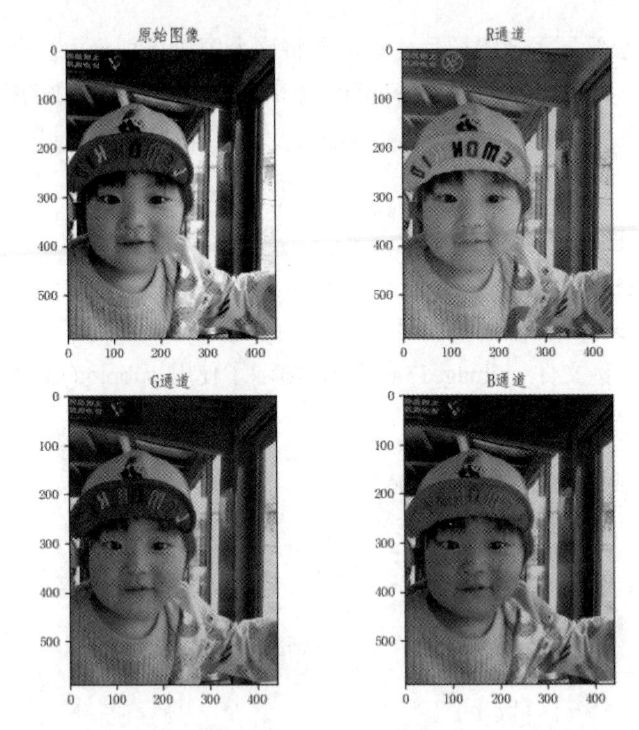

图 2-9　多个视图窗口

2.3.6　使用 subplots()函数绘制多视图窗口

在使用 scikit-image 绘制图片的过程中，我们可以用 subplots()函数来绘制多视图窗口。函数 subplots()分别返回一个窗口 figure 和一个 tuple 型的 ax 对象，该对象包含所有的子视图窗口。例如在下面的实例文件 skimage14.py 中，演示了使用 subplots()函数绘制多通道图像的过程。

源码路径：daima\2\2-12\skimage14.py

```
from skimage import data,io, color
import matplotlib.pyplot as plt
from pylab import mpl
#下面两行代码能保证汉字正确显示
mpl.rcParams['font.sans-serif'] = ['FangSong'] # 指定默认字体
mpl.rcParams['axes.unicode_minus'] = False # 解决保存图像是负号'-'显示为方块的问题
image = io.imread('111.jpg')
image_hsv = color.rgb2hsv(image)

fig, axes = plt.subplots(2, 2, figsize=(8, 8))
```

```
axe0, axe1, axe2, axe3 = axes.ravel()

axe0.imshow(image)
axe0.set_title('原始图像')

axe1.imshow(image_hsv[:, :, 0])
axe1.set_title('H 通道')

axe2.imshow(image_hsv[:, :, 1])
axe2.set_title('S 通道')

axe3.imshow(image_hsv[:, :, 2])
axe3.set_title('V 通道')

for ax in axes.ravel():
   ax.axis('off')

fig.tight_layout()

plt.show()
```

执行代码后不但显示原始图片，而且还会显示另外 3 个通道的子视图，如图 2-10 所示。

原始图像

H 通道

S 通道

V 通道

图 2-10　多个视图窗口

2.3.7 改变指定图片的大小

通过使用 scikit-image，我们可以对指定的图片实现缩放和旋转处理，这主要是通过其内置模块 transform 实现的。例如在下面的实例文件 skimage20.py 中，演示了使用函数 resize()改变指定图片大小的过程。

源码路径：daima\2\2-13\skimage20.py

```
from skimage import transform,data,io
import matplotlib.pyplot as plt
from pylab import mpl
#下面两行代码能保证汉字正确显示
mpl.rcParams['font.sans-serif'] = ['FangSong'] # 指定默认字体
mpl.rcParams['axes.unicode_minus'] = False # 解决保存图像是负号'-'显示为方块的问题
img = io.imread('111.jpg')
dst=transform.resize(img, (80, 60))
plt.figure('resize')
plt.subplot(121)
plt.title('原始图')
plt.imshow(img,plt.cm.gray)
plt.subplot(122)
plt.title('改变后')
plt.imshow(dst,plt.cm.gray)
plt.show()
```

通过上述代码，将图片 111.jpg 由原来的大小变成了 80×60 大小。执行代码后会通过两个子视图显示改变图像大小前后的对比，如图 2-11 所示。

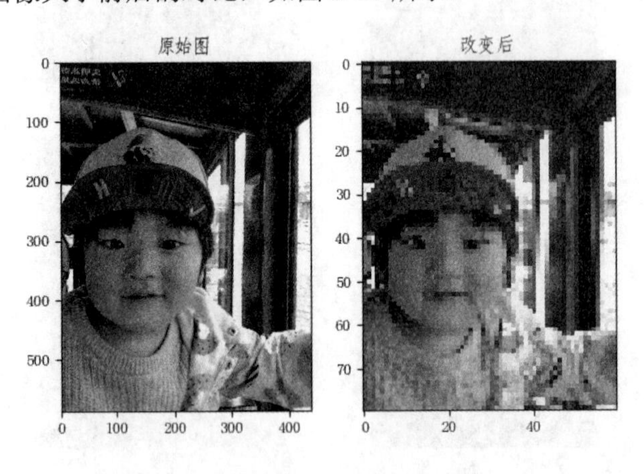

图 2-11 显示改变图像大小的前后对比

2.3.8 使用函数 rescale()缩放图片

在下面的实例文件 skimage21.py 中，演示了使用函数 rescale()缩放指定图片的过程。

源码路径：daima\2\2-14\skimage21.py

```
from skimage import transform,data,io
img = io.imread('111.jpg')
print(img.shape)                                #图片原始大小
print(transform.rescale(img, 0.1).shape)        #缩小为原来图片大小的 0.1 倍
print(transform.rescale(img, [0.5,0.25]).shape)
                                                #缩小为原来图片行数一半，列数四分之一
print(transform.rescale(img, 2).shape)          #放大为原来图片大小的 2 倍
```

执行代码后会显示不同缩放后的大小：

```
(588, 441, 3)
(59, 44, 3)
(294, 110, 3)
(1176, 882, 3)
```

2.3.9 使用函数 rotate()旋转图片

在下面的实例文件 skimage22.py 中，演示了使用函数 rotate()旋转指定图片的过程。

源码路径：daima\2\2-15\skimage22.py

```
from skimage import transform,io
import matplotlib.pyplot as plt
from pylab import mpl
#下面两行代码能保证汉字正确显示
mpl.rcParams['font.sans-serif'] = ['FangSong'] # 指定默认字体
mpl.rcParams['axes.unicode_minus'] = False # 解决保存图像是负号'-'显示为方块的问题
img=io.imread('111.jpg')
print(img.shape)                                #图片原始大小
img1=transform.rotate(img, 60)                  #旋转 60 度，不改变大小
print(img1.shape)
img2=transform.rotate(img, 30,resize=True)      #旋转 30 度，同时改变大小
print(img2.shape)
plt.figure('缩放')
plt.subplot(121)
plt.title('旋转 60 度')
plt.imshow(img1,plt.cm.gray)
```

```
plt.subplot(122)
plt.title('旋转 30 度')
plt.imshow(img2,plt.cm.gray)
plt.show()
```

执行代码后会输出显示原始图像大小、旋转 60 度时的大小和旋转 30 度时的大小：

```
(588, 441, 3)
(588, 441, 3)
(730, 676, 3)
```

并且还会分别显示旋转 60 度时和旋转 30 度时的图像效果，如图 2-12 所示。

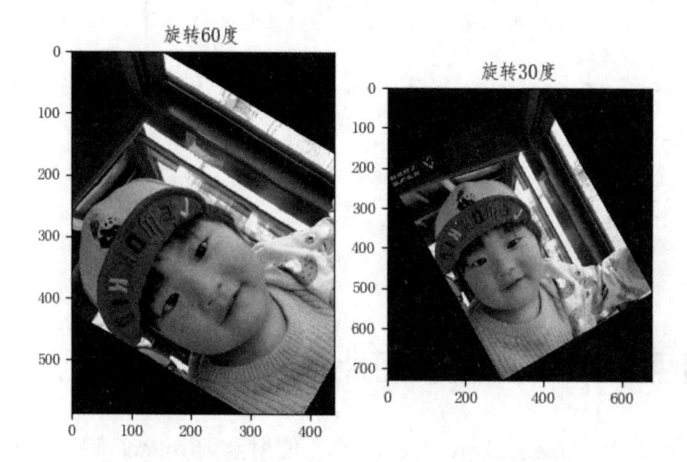

图 2-12　分别显示旋转 60 度时和旋转 30 度时的效果

第3章

OpenCV 图像视觉处理

OpenCV(Open Source Computer Vision Library)是一个开源的计算机视觉库，它提供了很多函数，这些函数非常高效地实现了计算机视觉算法(最基本的滤波到高级的物体检测皆有涵盖)。在本章的内容中，将详细讲解在 Python 程序中使用 OpenCV 实现图像视觉处理的知识，为读者步入本书后面知识的学习打下基础。

3.1 OpenCV 基础

OpenCV 是计算机视觉中经典的专用库，它支持多语言和跨平台，功能强大。为了让 Python 开发者使用 OpenCV 的强大功能，OpenCV 提供了 Python 接口库 OpenCV-Python，开发者通过调用 OpenCV-Python 中的成员模块和方法，从而可以在 Python 程序中使用 OpenCV 的强大功能。

扫码观看本节视频讲解

3.1.1 OpenCV 介绍

OpenCV 由 Gary Bradsky(加里·布拉德斯基)于 1999 年在英特尔创立。Gary Bradsky 当时在英特尔任职，怀着为计算机视觉和人工智能的从业者提供稳定的基础架构并以此来推动产业发展的美好愿景，启动了 OpenCV 项目。

OpenCV 支持多种编程语言，例如 C++、Python、Java 等，并且可在 Windows、Linux、OS X、Android 和 iOS 等不同平台上使用。OpenCV 的应用领域非常广泛，包括图像拼接、图像降噪、产品质检、人机交互、人脸识别、动作识别、动作跟踪、无人驾驶等。

OpenCV 的一个目标是提供易于使用的计算机视觉接口，从而帮助人们快速建立精巧的视觉应用。OpenCV 库包含从计算机视觉各个领域衍生出来的 500 多个函数，包括工业产品质量检验、医学图像处理、安保领域、交互操作、相机校正、双目视觉以及机器人学。

因为计算机视觉和机器学习经常在一起使用，所以 OpenCV 也包含一个完备的、具有通用性的机器学习库(ML 模块)。这个子库聚焦于统计模式识别以及聚类。ML 模块对 OpenCV 的核心任务(计算机视觉)相当有用，但是这个库也足够通用，可以用于任意机器学习问题。

3.1.2 OpenCV-Python 介绍

OpenCV-Python 是指解决计算机视觉问题的 Python 专用库。与 C/C++之类的语言相比，Python 语言的速度较慢。也就是说，可以使用 C/C++轻松扩展 Python，这使我们能够用 C/C++编写计算密集型代码并创建可用作 Python 模块的 Python 包装器。这会带来以下两个好处。

(1) 运行效率相差无几，因为实际上在后台运行的是通过 OpenCV-Python 调用的 C++代码。

(2) 在 Python 中比在 C/C++中更加容易地编写代码，OpenCV-Python 是原始 OpenCV C++实现的 Python 包装器。

OpenCV-Python 利用了 NumPy，这是一个高度优化的库，用于使用 MATLAB 样式的语法进行数值运算。所有 OpenCV 数组结构都与 NumPy 数组相互转换。这也使与使用 NumPy 的其他库(例如 SciPy 和 Matplotlib)的集成变得更加容易。

3.1.3 安装 OpenCV-Python

在 Windows 系统中安装 Python 后，可以使用如下 pip 命令安装 OpenCV-Python：

```
pip install opencv-python
```

本书中的内容只涉及了 OpenCV-Python，它只包含 OpenCV 的主要模块，这是完全免费的。其实还有一个库：opencv-contrib-python，这个库包含了 OpenCV 的主要模块以及扩展模块，扩展模块主要包含了一些带专利的收费算法(如 shift 特征检测)以及一些在测试的新的算法(稳定后会合并到主要模块)。在 Windows 系统中安装 Python 后，可以使用如下 pip 命令安装 opencv-contrib-python：

```
pip install opencv-contrib-python
```

3.2 OpenCV-Python 图像操作

在本节的内容中，首先学习读取图像、显示图像以及保存图像的知识，然后讲解绘图、图像算法和几何变换的知识。

扫码观看本节视频讲解

3.2.1 读取并显示图像

1. 读取图像

在 OpenCV-Python 中，使用内置函数 cv.imread()读取图像，被读取的图像应该在工作目录中，或使用图像的完整路径。函数 cv.imread()的语法格式如下：

```
cv.imread(filepath,flags)
```

参数说明如下。

- filepath：表示要读入图片的完整路径。

- flags：读入图片的标志，用于设置读取图像的方式，主要方式如下。
  - cv.IMREAD_COLOR：加载彩色图像，用整数 1 表示。任何图像的透明度都会被忽视，这是默认标志值。
  - cv.IMREAD_GRAYSCALE：以灰度模式加载图像，用整数 0 表示。
  - cv.IMREAD_UNCHANGED：加载图像，包括 alpha 通道，用-1 表示。

2. 显示图像

在 OpenCV-Python 中，使用内置函数 cv.imshow()可在窗口中显示图像，窗口会自动适合图像的尺寸大小。函数 cv.imshow()的语法格式如下：

```
cv.imshow(winname, mat)
```

参数说明如下。

- winname：表示窗口名称，它是一个字符串。
- mat：表示要显示的图像对象，我们可以根据需要创建任意多个窗口，也可以使用不同的窗口名称。

在下面的实例文件 cv01.py 中，演示了使用 OpenCV-Python 读取并显示指定图像的过程。

源码路径：daima\3\cv01.py

```
import cv2 as cv
print( cv.__version__ )
#加载彩色灰度图像
img = cv.imread('111.jpg',0)
cv.imshow('image',img)
cv.waitKey(0)
cv.destroyAllWindows()
```

代码说明如下。

- cv.__version__的功能是显示当前安装的 OpenCV-Python 版本。
- cv.imread('111.jpg',0)的功能是用灰度模式加载图片 111.jpg，最后通过函数 imshow('image',img)显示图片 111.jpg。
- cv.waitKey()是一个键盘绑定函数，其参数表示的时间以毫秒为单位。该函数等待任何键盘事件指定的毫秒。如果在这段时间内按下任何键，程序都将继续运行。如果**0**被传递，它将无限期地等待一次敲击键。它也可以设置为检测特定的按键，例如按下键 A 等。
- cv.destroyAllWindows()会销毁创建的所有窗口，如果要销毁任何特定的窗口，可使用函数 cv.destroyWindow()在其中传递确切的窗口名称作为参数。

执行效果如图 3-1 所示。

```
cv01 (1)
C:\ProgramData\Anaconda3\python.exe
4.2.0
```

图 3-1　执行效果

注意：在特殊情况下，可以创建一个空窗口，然后再将图像加载到该窗口。在这种情况下，可以指定窗口是否可调整大小，这是通过功能函数 cv.namedWindow()实现的。在默认情况下，该标志为 cv.WINDOW_AUTOSIZE。但是，如果将标志指定为 cv.WINDOW_NORMAL，则可以调整窗口大小。当图像尺寸过大以及向窗口添加跟踪栏时，这将很有帮助。例如下面的代码:

```
cv.namedWindow('image', cv.WINDOW_NORMAL)
cv.imshow('image', img)
cv.waitKey(0)
cv.destroyAllWindows()
```

3.2.2 保存图像

在 OpenCV-Python 中，使用内置函数 cv.imwrite()保存图像。例如：

```
cv.imwrite('messigray.png', img)
```

其中第一个参数表示需要保存图像的文件名，要保存图片为哪种格式，就带什么后缀。第二个参数表示保存的图像，该语句的功能是将图像以 PNG 格式保存在工作目录中。

例如在下面的实例 cv02.py 中，先准备一幅彩色图像文件 111.jpg。然后以灰度模式加载并显示图像文件 111.jpg，按 S 键后将灰度文件保存为 new.png 并退出，或者按 Esc 键直接退出而不保存。

源码路径：daima\3\cv02.py

```
import cv2 as cv
img = cv.imread('111.jpg',0)
cv.imshow('image',img)
k = cv.waitKey(0)
if k == 27:         # 等待Esc退出
    cv.destroyAllWindows()
elif k == ord('s'): # 等待关键字，保存和退出
    cv.imwrite('new.png',img)
    cv.destroyAllWindows()
```

执行效果如图3-2所示。

111.jpg

(a) 原来的彩色文件

new.png

(b) 保存后的灰度文件

图3-2　执行效果

3.2.3　在Matplotlib中显示图像

在Python程序中，经常使用Matplotlib实现绘图功能，例如数据可视化中的统计图。在使用OpenCV-Python显示图像时，可以借助于Matplotlib缩放图像或保存图像。例如在下面的实例cv03.py中，在Matplotlib中使用OpenCV-Python显示了一幅指定的图像。

源码路径：daima\3\cv03.py

```
import cv2 as cv
from matplotlib import pyplot as plt
img = cv.imread('111.jpg',0)
plt.imshow(img, cmap = 'gray', interpolation = 'bicubic')
plt.xticks([]), plt.yticks([])  # 隐藏 x 轴和 y 轴上的刻度值
plt.show()
```

执行效果如图 3-3 所示。

图 3-3　执行效果

3.2.4 绘图

在 OpenCV-Python 绘图应用中，经常用到的内置函数有 cv.line()、cv.circle()、cv.rectangle()、cv.ellipse()和 cv.putText()等，这些函数的常用参数如下。

- img：表示要绘制形状的图像。
- color：形状的颜色。对于 RGB 模式来说，将使用元组设置颜色，例如用(255,0,0)设置为蓝色。对于灰度模式来说，只需使用标量值设置即可。
- 厚度：线或圆等的粗细，默认值为 1。如果对闭合图形(如圆)传递-1，它将填充形状。
- lineType：线的类型，例如 8 连接线、抗锯齿线等。在默认情况下为 8 连接线。

(1) 画直线。

在 OpenCV-Python 中可使用内置函数 cv.line()绘制一条线，在绘制时需要设置开始坐标和结束坐标。例如在下面的实例 cv04.py 中，演示了在 OpenCV-Python 中绘制直线的过程。

源码路径：daima\3\cv04.py

```
import cv2 as cv
# 创建黑色的图像
img = cv.imread('111.jpg',1)
# 绘制一条厚度为 5 的蓝色对角线
cv.line(img,(0,0),(511,511),(255,0,0),5)
#显示图像
cv.imshow('image',img)
cv.waitKey(0)
```

在上述代码中，(0,0)表示起点，(511,511)表示终点，(255,0,0)表示线的颜色，5 表示线的宽度。执行效果如图 3-4 所示。

图 3-4　执行效果

(2) 画矩形。

在 OpenCV-Python 中可使用内置函数 cv.rectangle()绘制矩形，在绘制时需要设置矩形的左上角和右下角。请看下面的代码，将在图像 img 的右上角绘制一个绿色矩形。

```
cv.rectangle(img,(384,0),(510,128),(0,255,0),3)
```

(3) 画圆。

在 OpenCV-Python 中可使用内置函数 cv.circle()绘制圆，在绘制时需要设置圆的中心坐标和半径。例如：

```
cv.circle(img,(447,63), 63, (0,0,255), -1)
```

(4) 画椭圆。

在 OpenCV-Python 中可使用内置函数 cv.ellipse()绘制椭圆，在绘制时需要传递如下参数。

- 中心位置参数(x, y)。
- 轴长度参数(长轴长度, 短轴长度)。
- Angle 角度参数，是椭圆沿逆时针方向旋转的角度。startAngle 和 endAngle 分别表示从主轴沿顺时针方向测量的椭圆弧的开始和结束，若值为 0 和 360 则绘出完整的椭圆。

例如下面的代码绘制了一个椭圆：

```
cv.ellipse(img,(256,256),(100,50),0,0,180,255,-1)
```

(5) 画多边形。

在 OpenCV-Python 中可使用内置函数 cv.polylines()绘制多边形，在绘制时首先需要设置顶点坐标，将这些点组成形状为 ROWS*1*2 的数组，其中 ROWS 是顶点数，并且其类型应为 int32。例如下面绘制了一个带有 4 个顶点的黄色小多边形。

```
pts = np.array([[10,5],[20,30],[70,20],[50,10]], np.int32)
pts = pts.reshape((-1,1,2))
cv.polylines(img,[pts],True,(0,255,255))
```

注意：如果函数 cv.polylines()的第三个参数为 False，将获得一条连接所有点的折线，而不是闭合形状。

(6) 绘制文本。

在图像中绘制文本时需要设置多个内容参数，例如要写入的文字数据、要放置它的位置坐标(即数据开始的左下角)、字体类型、字体比例(指定字体大小)、颜色、厚度、线条类型等。为了获得更好的外观，建议使用 lineType = cv.LINE_AA。下面的代码将在白色图像上绘制文字“OpenCV”。

```
font = cv.FONT_HERSHEY_SIMPLEX
cv.putText(img,'OpenCV',(10,500), font, 4,(255,255,255),2,cv.LINE_AA)
```

请看下面的实例 cv05.py，我们将对上面的代码进行总结，演示使用 OpenCV-Python 绘制常见图形的过程。

源码路径：daima\3\cv05.py

```
import numpy as np
import cv2 as cv
# 创建黑色的图像
img = np.zeros((512,512,3), np.uint8)
# 绘制一条厚度为 5 的蓝色对角线
cv.line(img,(0,0),(511,511),(255,0,0),5)

cv.rectangle(img,(384,0),(510,128),(0,255,0),3)

cv.circle(img,(447,63), 63, (0,0,255), -1)

cv.ellipse(img,(256,256),(100,50),0,0,180,255,-1)

pts = np.array([[10,5],[20,30],[70,20],[50,10]], np.int32)
pts = pts.reshape((-1,1,2))
cv.polylines(img,[pts],True,(0,255,255))

font = cv.FONT_HERSHEY_SIMPLEX
cv.putText(img,'OpenCV',(10,500), font, 4,(255,255,255),2,cv.LINE_AA)

#显示图像
cv.imshow('image',img)
cv.waitKey(0)
```

执行效果如图 3-5 所示。

图 3-5 执行效果

3.2.5 将鼠标作为画笔

在使用电脑绘图时，通常将鼠标作为画笔。假设我们即将创建一个应用程序，要求双击鼠标后在图像上绘制一个圆。为了实现这个功能，首先创建一个鼠标回调函数，该函数在发生鼠标事件时执行。鼠标事件可以是与鼠标相关的任何事物，例如左键按下，右键按下，左键双击等。它为我们提供了每个鼠标事件的坐标(x, y)。通过此活动和地点，我们可以做任何喜欢的事情。通过如下 Python 代码，可以输出显示所有可用的鼠标事件。

```
import cv2 as cv
events = [i for i in dir(cv) if 'EVENT' in i]
print( events )
```

在作者的电脑中执行后会输出显示 OpenCV-Python 支持的鼠标事件：

```
['EVENT_FLAG_ALTKEY', 'EVENT_FLAG_CTRLKEY', 'EVENT_FLAG_LBUTTON',
'EVENT_FLAG_MBUTTON', 'EVENT_FLAG_RBUTTON', 'EVENT_FLAG_SHIFTKEY',
'EVENT_LBUTTONDBLCLK', 'EVENT_LBUTTONDOWN', 'EVENT_LBUTTONUP',
'EVENT_MBUTTONDBLCLK', 'EVENT_MBUTTONDOWN', 'EVENT_MBUTTONUP',
'EVENT_MOUSEHWHEEL', 'EVENT_MOUSEMOVE', 'EVENT_MOUSEWHEEL',
'EVENT_RBUTTONDBLCLK', 'EVENT_RBUTTONDOWN', 'EVENT_RBUTTONUP']
```

请看下面的实例 cv06.py，功能是创建一个矩形画布，当双击鼠标左键时调用函数 draw_circle()绘制一个指定样式的圆。

源码路径：daima\3\cv06.py

```
import numpy as np
import cv2 as cv
# 鼠标回调函数
def draw_circle(event,x,y,flags,param):
    if event == cv.EVENT_LBUTTONDBLCLK:
        cv.circle(img,(x,y),100,(255,0,0),-1)
# 创建一个黑色的图像，一个窗口，并绑定到窗口的功能
img = np.zeros((512,512,3), np.uint8)
cv.namedWindow('image')
cv.setMouseCallback('image',draw_circle)
while(1):
    cv.imshow('image',img)
    if cv.waitKey(20) & 0xFF == 27:
        break
cv.destroyAllWindows()
```

执行效果如图 3-6 所示。

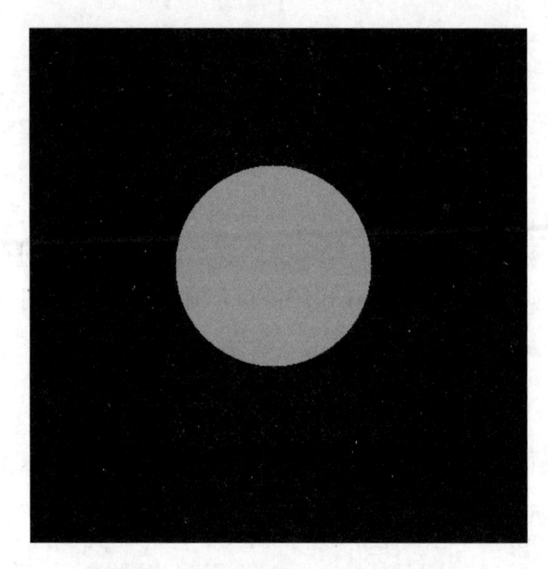

图 3-6　当双击鼠标左键时绘制圆

3.2.6　调色板程序

在 OpenCV-Python 中使用内置函数 cv.createTrackbar()创建滑块，例如：

```
cv.createTrackbar('R','image',0,255,nothing)
```

在函数 cv.createTrackbar()中，第一个参数表示滑块的名称，第二个参数是滑块附加到窗口的名称，第三个参数是默认值，第四个参数是最大值，第五个是执行的回调函数每次跟踪栏值更改。

在下面的实例中，使用内置函数 cv.createTrackbar()创建滑块，通过滑块可以设置屏幕的颜色。执行后先显示设置颜色窗口，以及 3 个用于指定 R、G、B 颜色的跟踪栏。通过滑动滑块可以相应地更改窗口的颜色。在默认情况下，初始颜色将设置为黑色。

源码路径：daima\3\cv07.py

```
import numpy as np
import cv2 as cv
def nothing(x):
   pass
# 创建一个黑色的图像，一个窗口
img = np.zeros((300,512,3), np.uint8)
cv.namedWindow('image')
# 创建颜色变化的轨迹栏
```

```
cv.createTrackbar('R','image',0,255,nothing)
cv.createTrackbar('G','image',0,255,nothing)
cv.createTrackbar('B','image',0,255,nothing)
# 为 ON/OFF 功能创建开关
switch = '0 : OFF \n1 : ON'
cv.createTrackbar(switch, 'image',0,1,nothing)
while(1):
   cv.imshow('image',img)
   k = cv.waitKey(1) & 0xFF
   if k == 27:
      break
   # 得到四条轨迹的当前位置
   r = cv.getTrackbarPos('R','image')
   g = cv.getTrackbarPos('G','image')
   b = cv.getTrackbarPos('B','image')
   s = cv.getTrackbarPos(switch,'image')
   if s == 0:
      img[:] = 0
   else:
      img[:] = [b,g,r]
cv.destroyAllWindows()
```

在上述代码中，我们创建了一个画笔开关，只有在该开关为 ON 的情况下，才能使用滑块设置屏幕的颜色，否则屏幕始终为黑色。执行效果如图 3-7 所示。

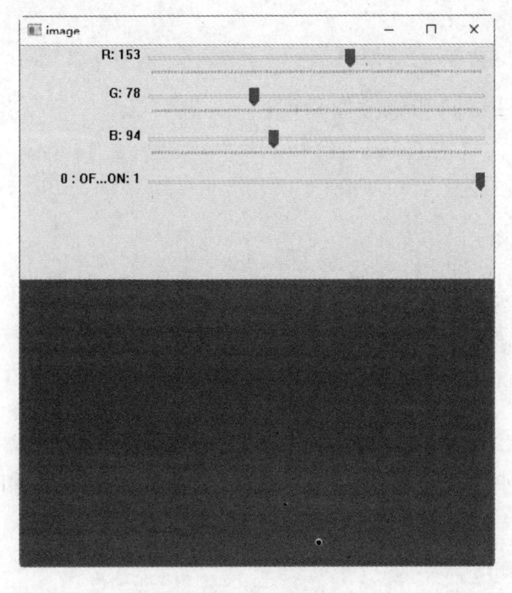

图 3-7　执行效果

3.2.7 基本的属性操作

(1) 访问和修改像素值。

首先加载一幅指定的彩色图像：

```
>>> import numpy as np
>>> import cv2 as cv
>>> img = cv.imread('111.jpg')
```

接下来可以通过行坐标和列坐标来访问图像的像素值。对于 RGB 图像来说，会返回一个由红色、绿色和蓝色值组成的数组。对于灰度图像来说，只会返回相应的灰度。

```
>>> px = img[100,100]
>>> print( px )
[157 166 200]
# 仅访问蓝色像素
>>> blue = img[100,100,0]
>>> print( blue )
157
```

也可以用相同的方式修改图像的像素值：

```
>>> img[100,100] = [255,255,255]
>>> print( img[100,100] )
[255 255 255]
```

也可以用下面的方法访问和修改像素值：

```
# 访问 RED 值
>>> img.item(10,10,2)
59
# 修改 RED 值
>>> img.itemset((10,10,2),100)
>>> img.item(10,10,2)
100
```

(2) 访问图像属性。

图像属性包括行数、列数、通道数、图像数据类型、像素数等。在 OpenCV-Python 中，可以通过 img.shape 访问图像的形状。img.shape 会返回由行、列和通道数组成的元组(如果图像是彩色的)，例如：

```
>>> print( img.shape )
(342, 548, 3)
```

可以通过访问 img.size 获得图像的像素总数：

```
>>> print( img.size )
562248
```

可以通过 img.dtype 获得图像的数据类型：

```
>>> print( img.dtype )
uint8
```

注意：如果图像是灰度模式，则返回的元组仅包含行数和列数，因此这是检查加载的图像是灰度还是彩色的好方法。

(3) 为图像设置边框(填充)。

如果要在图像周围创建边框(如相框)，可以使用 OpenCV-Python 内置函数 cv.copyMakeBorder()实现。函数 cv.copyMakeBorder()在卷积运算和零填充等方面的应用比较常见，此函数的常用参数如下。

- src：要处理的图像。
- top/bottom/left/right：边界的宽度和高度(以相应方向上的像素数为单位)。
- borderType：定义要添加哪种边框的标志，可以是以下类型。
  - cv.BORDER_CONSTANT：添加恒定的彩色边框。该值应作为下一个参数给出。
  - cv.BORDER_REFLECT：边框将是边框元素的镜像。
- value：边框的颜色。

请看下面的实例 cv08.py，演示了使用函数 cv.copyMakeBorder()为图像设置边框的过程。在 Matplotlib 中，为图像 111.jpg 设置了多种样式的边框。

源码路径：daima\3\cv08.py

```
import cv2 as cv
import numpy as np
from matplotlib import pyplot as plt
BLUE = [255,0,0]
img1 = cv.imread('111.jpg')
replicate = cv.copyMakeBorder(img1,10,10,10,10,cv.BORDER_REPLICATE)
reflect = cv.copyMakeBorder(img1,10,10,10,10,cv.BORDER_REFLECT)
reflect101 = cv.copyMakeBorder(img1,10,10,10,10,cv.BORDER_REFLECT_101)
wrap = cv.copyMakeBorder(img1,10,10,10,10,cv.BORDER_WRAP)
constant= cv.copyMakeBorder(img1,10,10,10,10,cv.BORDER_CONSTANT,value=BLUE)
plt.subplot(231),plt.imshow(img1,'gray'),plt.title('ORIGINAL')
plt.subplot(232),plt.imshow(replicate,'gray'),plt.title('REPLICATE')
plt.subplot(233),plt.imshow(reflect,'gray'),plt.title('REFLECT')
```

```
plt.subplot(234),plt.imshow(reflect101,'gray'),plt.title('REFLECT_101')
plt.subplot(235),plt.imshow(wrap,'gray'),plt.title('WRAP')
plt.subplot(236),plt.imshow(constant,'gray'),plt.title('CONSTANT')
plt.show()
```

执行效果如图 3-8 所示。

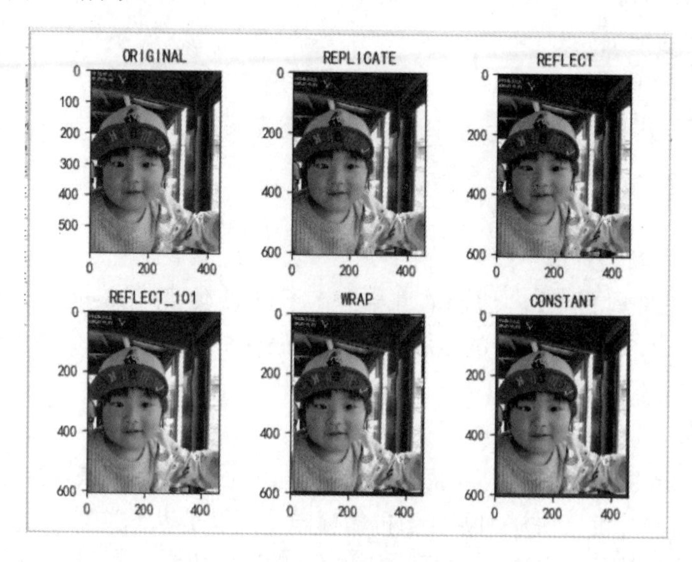

图 3-8　执行效果

3.2.8　图像的几何变换

(1)　缩放。

在 OpenCV-Python 中，使用内置函数 cv.resize()可实现图像的缩放功能。图像的缩放大小，既可以设置缩放比例手动设置，也可使用不同的插值方法。首选的插值方法是使用 cv.INTER_AREA 实现缩小，使用 cv.INTER_CUBIC 和 cv.INTER_LINEAR 实现缩放。在默认情况下，出于调整大小的目的，建议使用 cv.INTER_LINEAR。

请看下面的实例 cv09.py，功能是使用内置函数 cv.resize()将图像 111.jpg 放大两倍并显示出来。

源码路径：daima\3\cv09.py

```
import numpy as np
import cv2 as cv
img = cv.imread('111.jpg')
res = cv.resize(img,None,fx=2, fy=2, interpolation = cv.INTER_CUBIC)
```

```
#或者
height, width = img.shape[:2]
res = cv.resize(img,(2*width, 2*height), interpolation = cv.INTER_CUBIC)
cv.imshow('image',res)
cv.waitKey(0)
cv.destroyAllWindows()
```

执行效果如图 3-9 所示。

图 3-9 执行效果

(2) 平移。

平移是指移动物体的位置，如果想在(x,y)方向上移动，那么可以将移动轨迹设置为(t_x,t_y)，此时可以创建转换矩阵 M，则矩阵 M 的值为：

$$M=\begin{bmatrix}1 & 0 & t_x\\0 & 1 & t_y\end{bmatrix}$$

我们可以将转换矩阵 M 放入到 np.float32 类型的 NumPy 数组中，并将其传递给函数 cv.warpAffine()。请看下面的实例 cv10.py，功能是使用内置函数 cv.warpAffine()将图像 111.jpg 偏移为(100, 50)。

源码路径：daima\3\cv10.py

```
import numpy as np
import cv2 as cv
img = cv.imread('111.jpg',0)
rows,cols = img.shape
M = np.float32([[1,0,100],[0,1,50]])
dst = cv.warpAffine(img,M,(cols,rows))
cv.imshow('img',dst)
cv.waitKey(0)
cv.destroyAllWindows()
```

在上述代码中，函数 cv.warpAffine()的第三个参数用于设置图像的大小，设置格式为(width, height)，其中 width 为列数，height 为行数。执行效果如图 3-10 所示。

图 3-10 执行效果

(3) 旋转。

实际应用中，将图像旋转θ角度是通过以下形式的变换矩阵实现的：

$$M=\begin{bmatrix}\cos\theta & -\sin\theta \\ \sin\theta & \cos\theta\end{bmatrix}$$

但是在 OpenCV 中提供了可缩放的旋转以及可调整的旋转中心，因此开发者可以在自己喜欢的任何位置旋转。修改后的变换矩阵为：

$$\begin{bmatrix}\alpha & \beta & (1-\alpha)\cdot\text{center}\cdot x-\beta\cdot\text{center}\cdot y \\ -\beta & \alpha & \beta\cdot\text{center}\cdot x+(1-\alpha)\cdot\text{center}\cdot y\end{bmatrix}$$

其中：

α=scale · cosθ，β=scale · sinθ

为了找到实现上述转换矩阵的旋转功能，在 OpenCV 中提供了内置函数 cv.getRotationMatrix2D()。请看下面的实例 cv11.py，功能是将图像 111.jpg 相对于中心旋转 90 度而没有任何缩放比例。

源码路径：daima\3\cv11.py

```
import numpy as np
import cv2 as cv
img = cv.imread('111.jpg',0)
rows,cols = img.shape
M = np.float32([[1,0,100],[0,1,50]])
dst = cv.warpAffine(img,M,(cols,rows))
cv.imshow('img',dst)
cv.waitKey(0)
cv.destroyAllWindows()
```

执行效果如图 3-11 所示。

图 3-11　执行效果

3.2.9 图像直方图

将图像的像素组成转换成为直方图，可以总体了解图像的强度分布。在直方图的 *X* 轴上具有像素值(不总是从 0～255 的范围)，在直方图的 *Y* 轴上具有图像中相应像素数的图。通过查看图像的直方图，可以直观地了解该图像的对比度、亮度、强度分布等。在现实应用中，几乎所有图像处理工具都提供了直方图功能。在 OpenCV 和 NumPy 中都内置了直方图功能，在使用这些功能之前，需要先了解一些与直方图有关的技术术语。

- BINS：表示在直方图上面的显示每个像素值的像素数，即从 0～255。也就是说，我们需要用 256 个值来显示上面的直方图。但是考虑一下，如果不需要分别找到所有像素值的像素数，而是找到像素值间隔中的像素数怎么办？例如，需要找到介于 0～15 之间的像素数，然后找到 16～31 之间，…，240～255 之间的像素数。在这个时候，只需要 16 个值即可表示直方图。此时要做的是将整个直方图分成 16 个子部分，每个子部分的值就是其中所有像素数的总和。每个子部分都称为“BIN”。在第一种情况下，bin 的数量为 256 个(每个像素一个)，而在第二种情况下，bin 的数量仅为 16 个。在 OpenCV 中，用 histSize 术语表示 BINS。

- DIMS：表示为直方图收集数据的参数的数量。在这种情况下，因为仅收集关于强度值的数据，所以值是 1。
- RANGE：表示要测量的强度值的范围，通常是[0,256]，即所有强度值。

(1) OpenCV 中的直方图计算。

在 OpenCV 中，使用内置函数 cv.calcHist()计算直方图：

```
cv.calcHist(images, channels, mask, histSize, ranges [, hist [, accumulate]])
```

参数说明如下。

- images：是 uint8 或 float32 类型的源图像，应该被放在方括号中，即[img]。
- channels：也以方括号给出，表示计算直方图的通道的索引。例如，如果处理的是灰度图像，则其值为[0]。对于彩色图像来说，可以传递[0]、[1]或[2]分别计算红色、绿色或蓝色通道的直方图。
- mask：图像掩码。为了计算完整图像的直方图，将其指定为 None。但是，如果要计算某图像特定区域的直方图，则必须为此创建一个掩码图像并将其作为掩码。
- histSize：表示 BIN 计数，需要放在方括号中，全尺寸用[256]表示。
- ranges：表示 RANGE 等级，通常为[0,256]。
- hist：是 256*1 的数组，每个值对应于该图像中具有相应像素值的像素数。

请看下面的代码，功能是以灰度模式加载图像并得到其完整的直方图。

```
img = cv.imread('home.jpg',0)
hist = cv.calcHist([img],[0],None,[256],[0,256])
```

(2) NumPy 中的直方图计算。

在 NumPy 中提供内置函数 np.histogram()计算直方图，例如下面的代码：

```
hist,bins = np.histogram(img.ravel(),256,[0,256])
```

bin 具有 257 个元素，NumPy 计算出 bin 的范围为 0～0.99、1～1.99、2～2.99 等，所以

计算出的直方图的最终范围为 255～255.99。虽然在上述代码的最后添加了 256，但是其实不需要 256，最多 255 就足够了。

另外，在 NumPy 中还有另一个内置函数 np.bincount()，此函数的效率比 np.histogram() 快 10 倍左右。因此，建议使用函数 np.bincount()绘制一维直方图，但是不要忘记在 np.bincount()中设置 minlength = 256。例如“hist = np.bincount(img.ravel(), minlength = 256)”。

注意：OpenCV 函数比 np.histogram()快大约 40 倍，建议开发者尽可能使用 OpenCV 函数绘制直方图。

请看下面的实例 cv12.py，使用 Matplotlib 的内置函数 matplotlib.pyplot.hist()为图像 888.jpg 绘制了对应的直方图，而无须使用前面介绍的 calcHist()函数或 np.histogram()函数。

源码路径：daima\3\cv12.py

```
import cv2 as cv
from matplotlib import pyplot as plt
img = cv.imread('888.jpg',0)
plt.hist(img.ravel(),256,[0,256]); plt.show()
```

执行效果如图 3-12 所示。

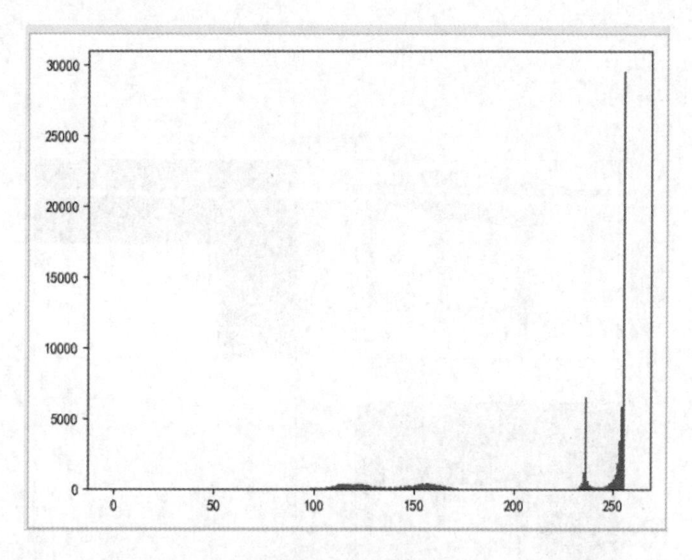

图 3-12 执行效果

在 OpenCV-Python 中可以调整直方图的值及其 bin 值，使其看起来像 x、y 坐标，以便可以进一步使用函数 cv.line()或函数 cv.polyline()绘制直方图。请看下面的实例 cv13.py，功能是使用 OpenCV-Python 函数 cv.calcHist()来绘制整个图像的直方图。如果只想绘制图像某些区域的直方图，应该如何实现呢？只需创建一个掩码图像即可，假设要找到直方图为白色，然后把这个作为掩码进行传递。

源码路径：daima\3\cv13.py

```
import numpy as np
import cv2 as cv
from matplotlib import pyplot as plt
img = cv.imread('888.jpg',0)

mask = np.zeros(img.shape[:2], np.uint8)
mask[100:300, 100:400] = 255
masked_img = cv.bitwise_and(img,img,mask = mask)
# 计算掩码区域和非掩码区域的直方图
# 检查作为掩码的第三个参数
hist_full = cv.calcHist([img],[0],None,[256],[0,256])
hist_mask = cv.calcHist([img],[0],mask,[256],[0,256])
plt.subplot(221), plt.imshow(img, 'gray')
plt.subplot(222), plt.imshow(mask,'gray')
plt.subplot(223), plt.imshow(masked_img, 'gray')
plt.subplot(224), plt.plot(hist_full), plt.plot(hist_mask)
plt.xlim([0,256])
plt.show()
```

执行效果如图 3-13 所示。

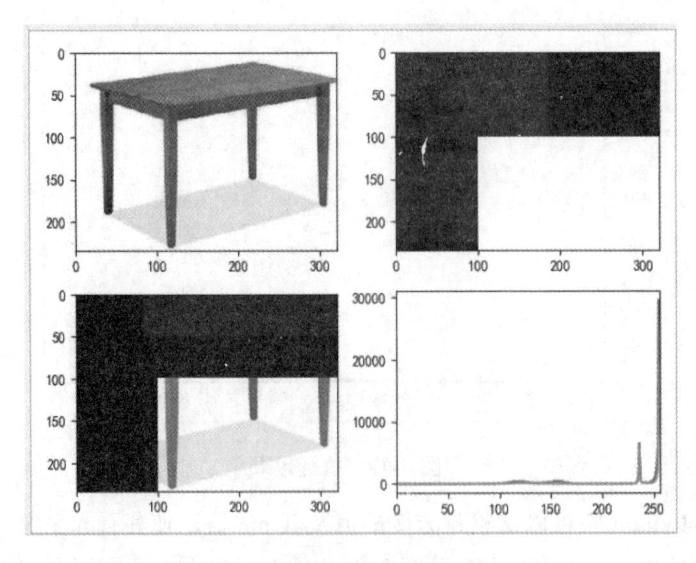

图 3-13　执行效果

3.2.10 特征识别：Harris(哈里斯)角检测

哈里斯角点是在任意方向上移动(u,v)，都会有很明显的变化。如图 3-14 所示，一个局部很小的区域，如果是在图片区域中移动灰度值没有变化，那么窗口内不存在角点。如果在某一个方向上移动，一侧发生很大变化而另一侧没有变化，那么说明这个区域是位于该对象的边缘区域。

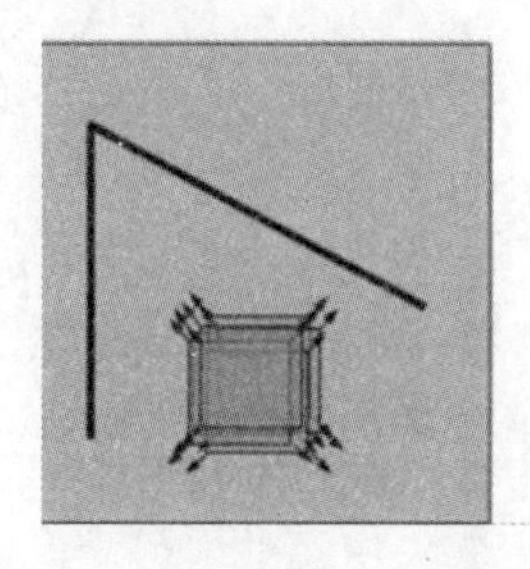
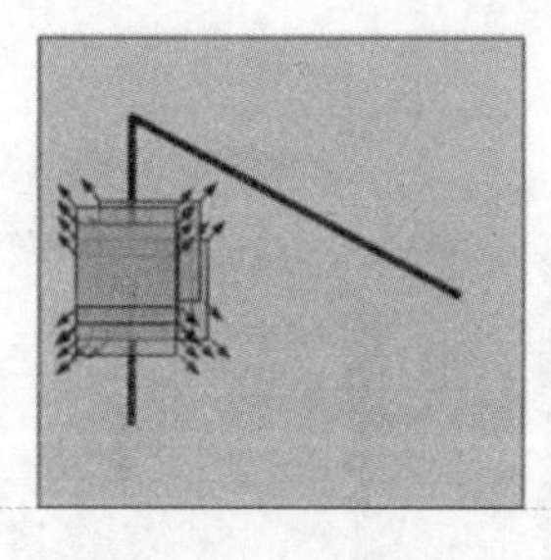
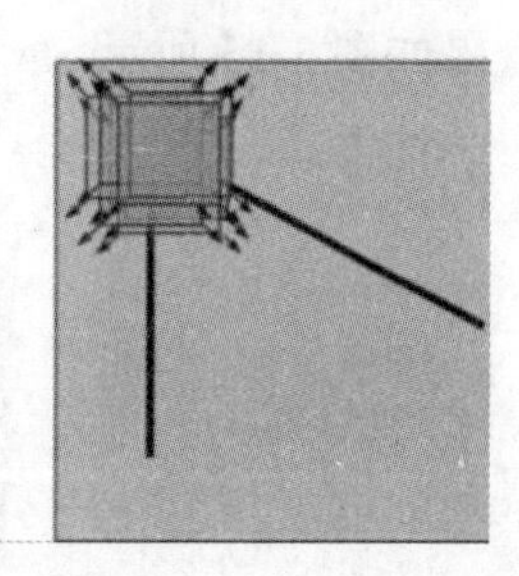

图 3-14 哈里斯角

在 OpenCV-Python 中提供了如下 Harris 角点检测函数：

```
cv2.cornerHarris(src, blockSize, ksize, k[, dst[,borderType]]) -> dst
```

参数说明如下。

- src：输入为单通道 8 位或者浮点数图片。
- dst：存储哈里斯角点检测响应的图像矩阵，矩阵大小跟 src 输入的一样，数据类型为浮点数。
- blockSize：领域大小。
- ksize：孔径参数。
- k：公式中的无限制参数。
- borderType：边界处理类型。

请看下面的实例 cv14.py，功能是使用 OpenCV-Python 函数 cv.cornerHarris()实现 Harris 角点检测。

源码路径：daima\3\cv14.py

```
import numpy as np
import cv2 as cv
filename = '888.jpg'
img = cv.imread(filename)
gray = cv.cvtColor(img,cv.COLOR_BGR2GRAY)
```

```
gray = np.float32(gray)
dst = cv.cornerHarris(gray,2,3,0.04)
#result 用于标记角点，并不重要
dst = cv.dilate(dst,None)
#最佳值的阈值，它可能因图像而异。
img[dst>0.01*dst.max()]=[0,0,255]
cv.imshow('dst',img)
if cv.waitKey(0) & 0xff == 27:
    cv.destroyAllWindows()
```

执行效果如图 3-15 所示。

图 3-15　执行效果

有时候，可能需要找到最精确的角落。在 OpenCV-Python 中提供了一个内置函数 cv.cornerSubPix()，此函数进一步细化了以亚像素精度检测到的角落。请看下面的实例 cv15.py，先找到 Harris 角，然后通过这些角的质心(在一个角上可能有一堆像素，取它们的质心)来细化它们。Harris 角用红色像素标记，精确角用绿色像素标记。在使用函数 cv.cornerSubPix()时，必须定义停止迭代的条件。我们在特定的迭代次数或达到一定的精度后停止它，还需要定义它将搜索角落的大小。

源码路径：daima\3\cv15.py

```
import numpy as np
import cv2 as cv
filename = '007.jpg'
img = cv.imread(filename)
gray = cv.cvtColor(img,cv.COLOR_BGR2GRAY)
# 寻找哈里斯角
gray = np.float32(gray)
dst = cv.cornerHarris(gray,2,3,0.04)
dst = cv.dilate(dst,None)
ret, dst = cv.threshold(dst,0.01*dst.max(),255,0)
dst = np.uint8(dst)
```

```
# 寻找质心
ret, labels, stats, centroids = cv.connectedComponentsWithStats(dst)
# 定义停止和完善拐角的条件
criteria = (cv.TERM_CRITERIA_EPS + cv.TERM_CRITERIA_MAX_ITER, 100, 0.001)
corners = cv.cornerSubPix(gray,np.float32(centroids),(5,5),(-1,-1),criteria)
# 绘制
res = np.hstack((centroids,corners))
res = np.int0(res)
img[res[:,1],res[:,0]]=[0,0,255]
img[res[:,3],res[:,2]] = [0,255,0]
cv.imwrite('subpixel5.jpg',img)
```

执行效果如图 3-16 所示。

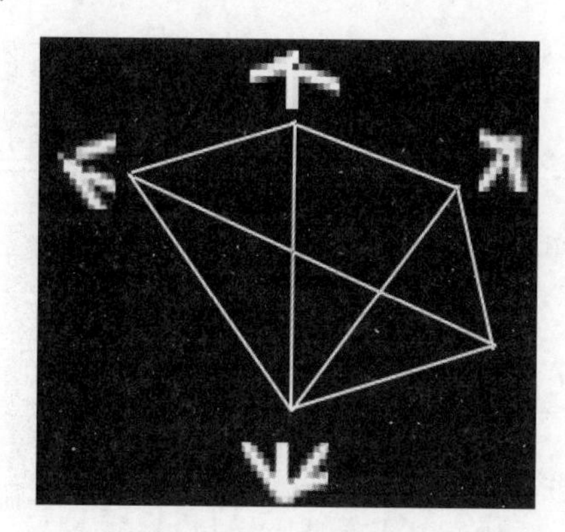

图 3-16 执行效果

3.3 OpenCV-Python 视频操作

在本节的内容中，首先学习读取视频、播放视频，以及保存视频的知识，然后讲解改变颜色空间和视频分析的知识。

扫码观看本节视频讲解

3.3.1 读取视频

在大多数情况下，需要使用摄像机捕捉实时画面，摄像机会提供一个非常简单的操作界面。请看下面的实例 cv16.py，功能是从摄像头中捕捉一段视频(例如使用笔记本电脑内置的网络摄像头)，并将其转换成灰度视频显示出来。要想捕获视频，需要创建一个 VideoCapture 对象，其参数可以是设备索引或视频文件的名称。设备索引是用于指定使用哪个摄像头的数字。在正常情况下，会连接并使用一个摄像头。在程序中通常传递 0(或-1)来调用一个摄像头，例如通过传递 0 来选择第 1 个相机，通过传递 1 来选择第 2 个相机，以此类推。在选择相机后可以逐帧捕获，在最后不要忘记释放捕获。

源码路径：daima\3\cv16.py

```
import numpy as np
import cv2 as cv
cap = cv.VideoCapture(0)
if not cap.isOpened():
   print("Cannot open camera")
   exit()
while True:
   # 逐帧捕获
   ret, frame = cap.read()
   # 如果正确读取帧，ret 为 True
   if not ret:
      print("Can't receive frame (stream end?). Exiting ...")
      break
   # 在框架上的操作出现在这里
   gray = cv.cvtColor(frame, cv.COLOR_BGR2GRAY)
   # 显示结果帧
   cv.imshow('frame', gray)
   if cv.waitKey(1) == ord('q'):
      break
# 完成所有操作后，释放捕获器
cap.release()
cv.destroyAllWindows()
```

执行代码后会打开当前电脑中的摄像头，并转换为灰度模式。执行效果如图 3-17 所示。

图 3-17 执行效果

在上述代码中，函数 cap.read()返回布尔值 True 或 False，如果正确读取了帧将返回

True。有时 cap 可能尚未初始化捕获，在这种情况下，此代码显示错误。可以通过函数 cap.isOpened()检查它是否已初始化，如果返回 True，那么说明已经初始化，否则使用 cap.open()打开。

另外，还可以使用函数 cap.get(propId)访问该视频的某些功能，其中 propId 是 0～18 之间的一个数字。每个数字表示视频的属性(如果适用于该视频)，并且可以通过 cv::VideoCapture::get()显示完整的详细信息，也可以使用 cap.set(propId, value)修改一些值，其中 value 是修改后的新值。

例如可以通过函数 cap.get(cv.CAP_PROP_FRAME_WIDTH)和 cap.get(cv.CAP_PROP_FRAME_HEIGHT)检查框架的宽度和高度。在默认情况下，它的分辨率为 640×480。但是如果想将其修改为 320×240，只需使用 and 运算符即可，例如：

```
ret = cap.set(cv.CAP_PROP_FRAME_WIDTH,320) and ret = cap.set(cv.CAP_PROP_
FRAME_HEIGHT,240)
```

3.3.2 播放视频

OpenCV-Python 播放视频的方法与从相机读取视频的方法相似，只是需要提前准备好要播放的视频。另外，在显示视频时，需设置适当的时间 cv.waitKey()。如果太小，则播放视频的速度会非常快，而如果太大，则播放视频的速度会很慢(显示慢动作)。在正常情况下，设置为 25 毫秒。请看下面的实例 cv17.py，功能是播放预先准备好的视频文件 capture-1.mp4。

源码路径：daima\3\cv17.py

```
import cv2 as cv
cap = cv.VideoCapture('capture-1.mp4')
while cap.isOpened():
    ret, frame = cap.read()
    # 如果正确读取帧，ret 为 True
    if not ret:
        print("Can't receive frame (stream end?). Exiting ...")
        break
    gray = cv.cvtColor(frame, cv.COLOR_BGR2GRAY)
    cv.imshow('frame', gray)
    if cv.waitKey(1) == ord('q'):
        break
cap.release()
cv.destroyAllWindows()
```

执行效果如图 3-18 所示。

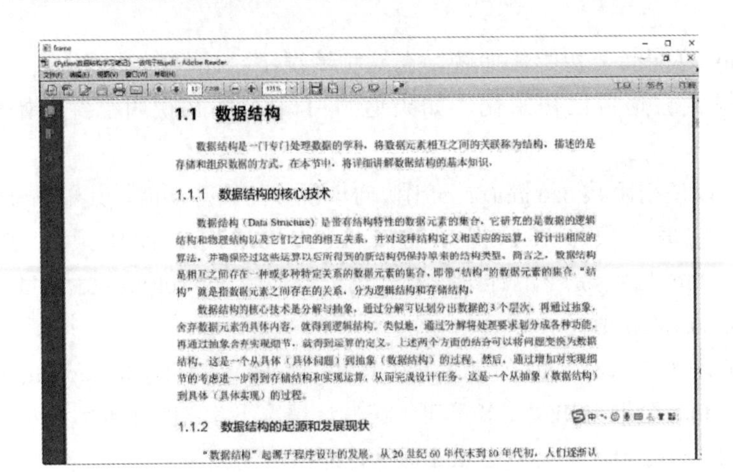

1.1 数据结构

数据结构是一门专门处理数据的学科，将数据元素相互之间的关联称为结构，描述的是存储和组织数据的方式。在本节中，将详细讲解数据结构的基本知识。

1.1.1 数据结构的核心技术

数据结构（Data Structure）是带有结构特性的数据元素的集合，它研究的是数据的逻辑结构和物理结构以及它们之间的相互关系，并对这种结构定义相适应的运算，设计出相应的算法，并确保经过这些运算以后所得到的新结构仍保持原来的结构类型。简言之，数据结构是相互之间存在一种或多种特定关系的数据元素的集合，即带"结构"的数据元素的集合。"结构"就是指数据元素之间存在的关系，分为逻辑结构和存储结构。

数据结构的核心技术是分解与抽象。通过分解可以划分出数据的3个层次，再通过抽象，舍弃数据元素的具体内容，就得到逻辑结构。类似地，通过分解将处理要求划分成各种功能，再通过抽象舍弃实现细节，就得到运算的定义。上述两个方面的结合可以将问题变换为数据结构。这是一个从具体（具体问题）到抽象（数据结构）的过程。然后，通过增加对实现细节的考虑进一步得到存储结构和实现运算，从而完成设计任务。这是一个从抽象（数据结构）到具体（具体实现）的过程。

1.1.2 数据结构的起源和发展现状

"数据结构"起源于程序设计的发展。从20世纪60年代末到80年代初，人们逐渐认

图 3-18　执行效果

3.3.3 保存视频

在捕捉读取一个视频时，可以一帧一帧地读取处理，然后可以使用函数 cv.imwrite()将读取到的图像帧保存起来。在保存视频时，先创建一个 VideoWriter 对象，并设置保存视频的文件名(例如 output.avi)，然后设置帧率的数量和帧的大小。最后设置颜色标志，如果设置为True，则使用编码器期望颜色帧保存，否则使用灰度帧保存。请看下面的实例cv18.py，功能是读取摄像头中的视频，并保存为视频文件 output.avi。

源码路径：daima\3\cv18.py

```
import cv2 as cv
cap = cv.VideoCapture(0)
# 定义编解码器并创建 VideoWriter 对象
fourcc = cv.VideoWriter_fourcc(*'XVID')
out = cv.VideoWriter('output.avi', fourcc, 20.0, (640,  480))
while cap.isOpened():
    ret, frame = cap.read()
    if not ret:
        print("Can't receive frame (stream end?). Exiting ...")
        break
    frame = cv.flip(frame, 0)
    # 写翻转的框架
    out.write(frame)
    cv.imshow('frame', frame)
    if cv.waitKey(1) == ord('q'):
        break
# 完成工作后释放所有内容
```

```
cap.release()
out.release()
cv.destroyAllWindows()
```

执行代码后会打开摄像头录制视频，按 Q 键后停止录制，并将录制的视频保存为output.avi。执行效果如图 3-19 所示。

图 3-19 执行效果

3.3.4 改变颜色空间

在 OpenCV 中大约有超过 150 种实现颜色空间转换的方法，但是最常用的方法有两种：BGR↔灰度和 BGR↔HSV。在 OpenCV-Python 中，使用内置函数 cv.cvtColor(input_image, flag) 改变颜色空间，其中参数 flag 用于设置转换的类型。

- 对于 BGR→灰度转换来说，使用 cv.COLOR_BGR2GRAY。
- 对于 BGR→HSV 转换来说，使用 cv.COLOR_BGR2HSV。

通过下面的代码，可以获取参数 flag 的其他转换类型：

```
>>> import cv2 as cv
>>> flags = [i for i in dir(cv) if i.startswith('COLOR_')]
>>> print( flags )
```

大家需要注意的是，HSV 的色相范围为[0,179]，饱和度范围为[0,255]，值范围为[0,255]。不同的软件使用不同的范围。因此，如果要将 OpenCV 值和它们进行比较，需要标准化处理这些范围。请看下面的实例 cv19.py，演示了使用内置函数 cv.cvtColor()改变摄像头视频颜色空间的知识。

源码路径：daima\3\cv19.py

```
import numpy as np
import cv2 as cv
img = cv.imread('111.jpg',0)
rows,cols = img.shape
# cols-1 和 rows-1 是坐标限制
M = cv.getRotationMatrix2D(((cols-1)/2.0,(rows-1)/2.0),90,1)
dst = cv.warpAffine(img,M,(cols,rows))
cv.imshow('img',dst)
cv.waitKey(0)
cv.destroyAllWindows()
```

执行效果如图 3-20 所示。

图 3-20　执行效果

3.3.5　视频的背景分离

背景分离(BS)是一种通过使用静态相机来生成前景掩码(即包含属于场景中的移动对象像素的二进制图像)的技术。也就是说，BS 用于计算前景掩码，在当前帧与背景模型之间执行减法运算，其中包含场景的静态部分，或者更一般而言，考虑到所观察场景的特征，可以将其视为背景的所有内容。请看如图 3-21 所示的背景分离，将视频中的一艘船和背景分离出来。

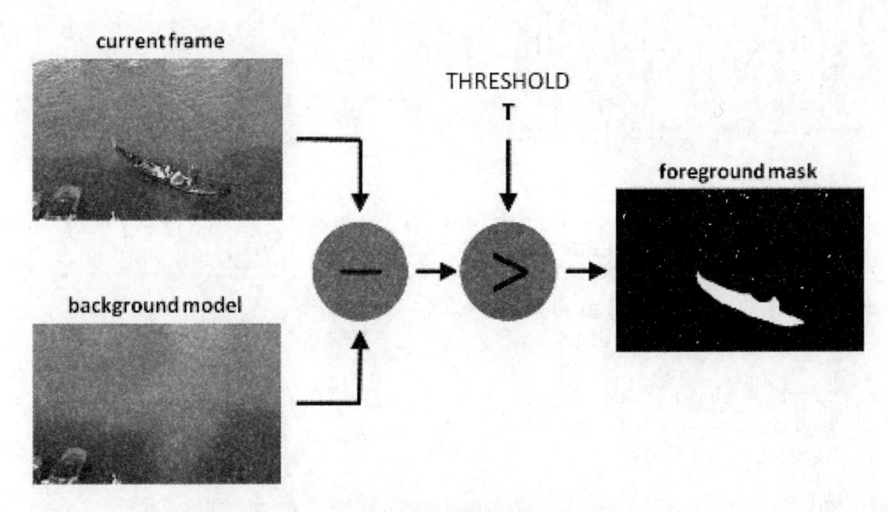

图 3-21 背景分离

在 OpenCV-Python 中实现背景分离的流程是，首先使用 cv.VideoCapture()从视频或图像序列中读取数据，然后使用类 cv.BackgroundSubtractor 创建和更新背景类，最后使用 cv.imshow 获取并显示前景蒙版。

请看下面的实例 cv20.py，可以让用户选择处理视频文件或图像序列，使用函数 cv.createBackgroundSubtractorMOG2()生成前景掩码。

源码路径：daima\3\cv20.py

```
import cv2 as cv
import argparse
parser = argparse.ArgumentParser(description='This program shows how to use
                              background subtraction methods provided by \
                              OpenCV. You can process both videos and images.')
parser.add_argument('--input', type=str, help='Path to a video or a sequence
of image.', default='capture-1.mp4')
parser.add_argument('--algo', type=str, help='Background subtraction method
                    (KNN, MOG2).', default='MOG2')
args = parser.parse_args()
if args.algo == 'MOG2':
   backSub = cv.createBackgroundSubtractorMOG2()
else:
   backSub = cv.createBackgroundSubtractorKNN()
capture = cv.VideoCapture(cv.samples.findFileOrKeep(args.input))
if not capture.isOpened:
   print('Unable to open: ' + args.input)
   exit(0)
while True:
```

```
    ret, frame = capture.read()
    if frame is None:
        break
    fgMask = backSub.apply(frame)

    cv.rectangle(frame, (10, 2), (100,20), (255,255,255), -1)
    cv.putText(frame, str(capture.get(cv.CAP_PROP_POS_FRAMES)), (15, 15),
               cv.FONT_HERSHEY_SIMPLEX, 0.5 , (0,0,0))

    cv.imshow('Frame', frame)
    cv.imshow('FG Mask', fgMask)

    keyboard = cv.waitKey(30)
    if keyboard == 'q' or keyboard == 27:
        break
```

代码说明如下。

- 函数 cv.createBackgroundSubtractor()用于生成前景掩码。在本实例中使用了默认参数，但是也可以在函数 create()中声明特定的参数。
- 函数 cv.VideoCapture()用于读取输入视频或输入图像序列。
- 通过代码 fgMask = backSub.apply(frame)更新背景模型，每一帧都用于计算前景掩码和更新背景。如果要更改用于更新背景模型的学习率，可以通过将参数传递给函数 apply()的方法来设置特定的学习率。
- 使用函数 cv.VideoCapture()提取当前帧号，并标记在当前帧的左上角，再使用白色矩形突出显示黑色的帧编号。

执行效果如图 3-22 所示。

图 3-22　执行效果

3.4 简易车牌识别系统

经过本书前面内容的学习，读者已经了解了在 Python 程序中使用 OpenCV-Python 处理图像和视频的知识。在本节的内容中，将通过一个车牌识别系统，介绍在商业项目中使用 OpenCV-Python 的方法。

扫码观看本节视频讲解

3.4.1 系统介绍

在本实例中预先准备几幅车牌图片，然后使用 scikit-image 和 OpenCV-Python 识别图片中的车牌号，并将识别的车牌号保存到 JSON 文件中。本实例主程序文件是 main.py，使用 Python 命令运行文件 main.py 的格式如下：

```
python main.py images_dir results_file
```

在上述命令中，images_dir 表示要识别的车牌照片目录，results_file 表示保存识别结果的 JSON 文件名。

在本项目中进行了如下设置。

- 设置车牌上有 7 个字符。
- 车牌的宽度大于图像宽度的 1/3，车牌的倾斜度不大于 45 度。
- 识别每个图像的最大处理时间为 2 秒。

3.4.2 通用程序

编写程序文件 utils.py，在里面定义了本实例用到的通用程序，由多个功能函数构成。

(1) 编写函数 sort_cornes(corns, img)，功能是实现透视变换，并实现变换后的排序[ul, ur, bl, br]。透视变换(Perspective Transformation)的本质是将图像投影到一个新的视平面，本函数能够计算车牌角和整个图像角之间的距离，然后找出[ul, ur, bl, br]中的内容。在得到排序列表[ul, ur, bl, br]后，我们就可以从图像中裁剪出车牌号。

函数 sort_cornes(corns, img)的代码如下：

```
def sort_cornes(corns, img):
    sorted_arr = []
    image_corns = []
    upper_right = [img.shape[1], 0]
    upper_left = [0, 0]
```

```
    bottom_right = [img.shape[1], img.shape[0]]
    bottom_left = [0, img.shape[0]]
    image_corns.append(upper_left)
    image_corns.append(upper_right)
    image_corns.append(bottom_left)
    image_corns.append(bottom_right)
    order = []
    ord = 0

    #计算车牌角和整个图像角之间的距离，找出[ul,ur,bl,br]中的字符
    for ind, val in enumerate(image_corns):
        lowest_dist = sqrt(img.shape[1]**2 + img.shape[0]**2)
        for i, v in enumerate(corns):
            dist = sqrt((val[0] - v[0])**2 + (val[1] - v[1])**2)
            if dist < lowest_dist:
                lowest_dist = dist
                ord = i
        order.append(ord)

    for o in order:
        sorted_arr.append(corns[o])

    # 一旦有了一个排序列表，就可以从图像中裁剪出车牌号
    return perspective(sorted_arr, img)
```

(2) 编写函数 perspective()，功能是从指定的图像中裁剪车牌。代码如下：

```
def perspective(arr, img, width=1000, height=250):
    p1 = np.float32([arr[0], arr[1], arr[2], arr[3]])
    p2 = np.float32([[0, 0], [width, 0], [0, height], [width, height]])

    get_tf = cv2.getPerspectiveTransform(p1, p2)
    persp = cv2.warpPerspective(img, get_tf, (width, height))

    return persp
```

(3) 编写函数 div_func()，功能是对裁剪的车牌实现颜色转换。代码如下：

```
def div_func(image, th=115):
    img_gray = cv2.cvtColor(image, cv2.COLOR_BGR2GRAY)
    blurred = cv2.GaussianBlur(img_gray, (5, 5), 0)
    _, thresh = cv2.threshold(blurred, th, 255, cv2.THRESH_BINARY_INV)

    contours, _ = cv2.findContours(thresh, cv2.RETR_EXTERNAL,
                                   cv2.CHAIN_APPROX_SIMPLE)
    contours = sorted(contours, key=cv2.contourArea, reverse=True)
    boxes = []
```

```
    for c in contours:
        (x, y, w, h) = cv2.boundingRect(c)
        #检查高度和宽度是否合理
        if h > 150 and (w > 15 and w < 200):
            cv2.rectangle(image, (x, y), (x + w, y + h), (0, 0, 0), 1)
            boxes.append([[x, y], [x + w, y], [x, y + h], [x + w, y + h]])
    return boxes
```

(4) 编写函数 divide_tab()，功能是用数字和字母在线段上划分裁剪区域，返回包含字母和数字的列表。代码如下：

```
def divide_tab(image, th=115):
    temp_img = image.copy()
    bboxes = div_func(image, th)

    if len(bboxes) < 7:
        th += 10
        if th >= 30:
            bboxes = div_func(temp_img, th)

    bboxes.sort()
    num_let_arr = []

    for b in bboxes:
        # 裁剪的单个值
        num_let = perspective(b, image, width=120, height=220)
        num_let_arr.append(num_let)

    # 返回包含字母和数字的列表
    return num_let_arr
```

(5) 编写函数 recognize()，功能是识别裁剪区域中的每个字母和数字。在识别每个字母时，会参考图像字符的概率(所有字母和数字)，找到可能性概率最大的字母并将其附加到列表中。代码如下：

```
def recognize(letters, im_paths):

    reference = []
    names = []
    # 读取参考图像并将其添加到数组(同时添加名称)
    for image_path in im_paths:
        image = cv2.imread(str(image_path), 0)
        reference.append(image)
        names.append(image_path.name[:1])

    text_arr = []
```

```
    #对于每个字母，计算参考图像的概率(所有字母和数字)
    for let in letters:
        probabilites = []
        letters_arr = []
        img_gray = cv2.cvtColor(let, cv2.COLOR_BGR2GRAY)
        blurred = cv2.GaussianBlur(img_gray, (3, 3), 0)
        ret, thresh = cv2.threshold(blurred, 135, 255, cv2.THRESH_BINARY)

        for ind, im in enumerate(reference):
            im = cv2.resize(im, (thresh.shape[1], thresh.shape[0]))
            result, _ = compare_ssim(thresh, im, full=True)
            letter = names[ind]
            probabilites.append(result)
            letters_arr.append(letter)

        # 找到概率最大的字母并将其附加到列表中
        max_val = max(probabilites)

        for ind, val in enumerate(probabilites):
            if max_val == val:
                if letters_arr[ind] != 'w' and letters_arr[ind] != 'c' and
                                        letters_arr[ind] != 'r':
                    text_arr.append(letters_arr[ind])
    return ''.join(text_arr)
```

(6) 编写函数 help_perform()，功能是对指定图像实现颜色转换，检查每个部分是否有4个角，然后计算角之间的距离，并除以这些值，以确定这是否是一个车牌号。代码如下：

```
def help_perform(image, th=135, wind=3):
    wrapped_tab = None
    img_gray = cv2.cvtColor(image, cv2.COLOR_BGR2GRAY)
    w = image.shape[1]
    h = image.shape[0]
    blurred = cv2.GaussianBlur(img_gray, (wind, wind), 3)
    _, thresh = cv2.threshold(blurred, th, 255, cv2.THRESH_BINARY_INV)

    cont, _ = cv2.findContours(thresh, mode=cv2.RETR_TREE,
                              method=cv2.CHAIN_APPROX_NONE)
    hull_list = []

    for i in range(len(cont)):
        h = cv2.convexHull(cont[i])
        hull_list.append(h)

    contours = sorted(hull_list, key=cv2.contourArea, reverse=True)[:8]
```

```
    for c in contours:
        clos = cv2.arcLength(c, True)
        apr = cv2.approxPolyDP(c, 0.05 * clos, True)
        cnt = 0
        nums = []
        points_arr = []

        # Check if c has 4 corners
        if len(apr) == 4:
            for i in range(4):
                j = i + 1
                if j == 4:
                    j = 0
                k = i - 1
                if k == -1:
                    k = 3

                # 计算角点之间的距离，然后除以这些值，以确定这是否是一个车牌号
                w_len_1 = sqrt((apr[i][0][0] - apr[j][0][0]) ** 2 + (apr[i][0][1]
                               - apr[j][0][1]) ** 2)
                w_len_2 = sqrt((apr[i][0][0] - apr[k][0][0]) ** 2 + (apr[i][0][1]
                               - apr[k][0][1]) ** 2)

                if w_len_1 > w_len_2:
                    if w_len_1 / w_len_2 < 3 or w_len_1 / w_len_2 > 7.3 or w_len_1
                                >= w:
                        continue
                else:
                    if w_len_2 / w_len_1 < 3 or w_len_2 / w_len_1 > 7.3 or w_len_2
                                >= w:
                        continue

                # 下一个判断条件是确定这是否是车牌号
                if w_len_1 >= w / 3 or w_len_2 >= w / 3:
                    cnt += 1

                    points_arr.append(apr[i][0])
                if cnt == 2:
                    nums.append(apr)

        if len(nums) == 1:
            wrapped_tab = sort_cornes(points_arr, image)

    return wrapped_tab
```

(7) 编写函数 perform_processing()，功能是调用上面的功能函数实现读取图像和裁剪、识别等处理功能。如果没有识别出任何内容，则返回问号???????。代码如下：

```
def perform_processing(image: np.ndarray, ref) -> str:
    th = 135
    win = 3
    wrapped_tab = None
    for i in range(5):
        wrapped_tab = help_perform(image, th=th, wind=win)
        th -= 5
        win += 6
        if wrapped_tab is not None:
            break
    # 如果没有找到任何内容，请返回问号
    if wrapped_tab is None:
        return '???????'
    try:
        single = divide_tab(wrapped_tab, th=115)
        resu = recognize(single, ref)
        if len(resu) == 7:
            return resu
        elif len(resu) > 7:
            x = len(resu) - 7
            return resu[x:]
        elif len(resu) == 0:
            return '???????'
        elif 7 > len(resu) > 0:
            x = 7 - len(resu)
            for i in range(x):
                resu = '?' + resu[0:]
            return resu
    except NameError:
        return '???????'

    return '???????'
```

3.4.3 主程序

编写本项目的主程序文件 main.py，功能是调用文件 utils.py 中的功能函数，实现车牌识别功能。首先使用函数 add_argument()添加两个 Python 命令参数 images_dir 和 results_file，然后根据用户设置的参数 images_dir 逐一读取 images_dir 目录中的图片并实现车牌识别，并将识别结果保存为 results_file。文件 main.py 的具体实现代码如下。

```
from processing.utils import perform_processing

def main():
    parser = argparse.ArgumentParser()
```

```
    parser.add_argument('images_dir', type=str)
    parser.add_argument('results_file', type=str)
    args = parser.parse_args()

    im_dir = Path('./numbers_letters/')
    im_paths = sorted([im_path for im_path in im_dir.iterdir() if
                       im_path.name.endswith('.jpeg')])

    images_dir = Path(args.images_dir)
    results_file = Path(args.results_file)

    images_paths = sorted([image_path for image_path in images_dir.iterdir() if
                           image_path.name.endswith('.jpg')])
    results = {}
    for image_path in images_paths:
        image = cv2.imread(str(image_path))
        if image is None:
            print(f'Error loading image {image_path}')
            continue

        results[image_path.name] = perform_processing(image, im_paths)
    print(os.path.abspath(images_dir))
    print(os.path.abspath(results_file))
    with results_file.open('w') as output_file:
        json.dump(results, output_file, indent=4)

if __name__ == '__main__':
    main()
```

假设在程序根目录 car 中保存了一幅图片，如图 3-23 所示。那么可以通过如下命令实现图片识别：

```
python main.py car 789.json
```

图 3-23 要识别的图片

运行上述命令后，会分析目录 car 中的所有图片，并识别提取图片里面的车牌号，然后将识别的车牌号保存在文件 789.json 中，如图 3-24 所示。

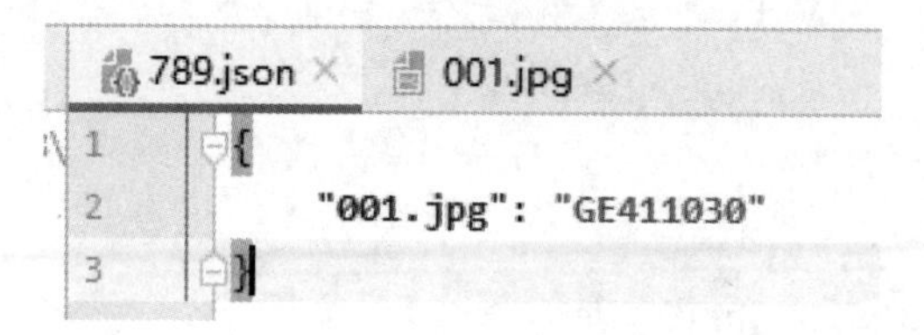

图 3-24　识别结果

第 4 章

dlib 机器学习和图像处理算法

dlib 是一个包含机器学习算法的 C++开源工具包，可以帮助开发者编写机器学习程序。目前 dlib 已经被广泛地应用在开发和学术领域，包括机器人、嵌入式设备、移动电话和大型高性能计算环境。在本章的内容中，将详细讲解 dlib 机器学习和图像处理算法的知识，为读者步入本书后面知识的学习打下基础。

4.1 dlib 介绍

扫码观看本节视频讲解

dlib 是由 C++语言编写的一个第三方库，提供了和机器学习、数值计算、图模型算法、图像处理等领域相关的一系列功能。库 dlib 的主要特点如下。

(1) 文档齐全：dlib 为每一个类和函数提供了完整的文档说明，同时还提供了 debug(调试)模式。在打开 debug 模式后，开发者在调试代码的过程中可以及时查看变量和对象的值，快速找出错误点。另外，dlib 还提供了大量的实践实例。

(2) 高质量的可移植代码：dlib 不依赖第三方库，可完美运行在 Windows、Mac OS 和 Linux 系统上。

(3) 提供了大量的机器学习和图像处理算法：

- 深度学习算法。
- 聚类分析和算法。
- 基于 SVM 的分类和递归算法。
- 针对大规模分类和递归的降维方法。
- 相关向量机(Relevance Vector Machine)，这是与支持向量机相同的函数形式稀疏概率模型，对未知函数进行预测或分类。具体训练是在贝叶斯框架下进行的，与 SVM 相比，不需要估计正则化参数，其核函数也不需要满足 Mercer 条件，需要更少的相关向量，训练时间长，测试时间短。

在使用 dlib 之前，需要先使用如下命令安装库文件：

```
pip install dlib
```

4.2 dlib 基本的人脸检测

扫码观看本节视频讲解

通过使用 dlib 提供的内置类和函数，可以方便地检测各种素材图像。在本节的内容中，将详细讲解使用 dlib 实现基本人脸检测的知识。

4.2.1 人脸检测

请看下面的实例文件 first.py，演示了使用 dlib 实现简易人脸检测的过程。

源码路径：daima\4\first.py

(1) 首先使用 import 语句导入 dlib 模块，代码如下：

```
import dlib
from imageio import imread
import glob
```

(2) 准备人脸检测器和显示窗口，设置要检测的图片路径是 111.jpg，在检测时指定一个阈值。代码如下：

```
path = '111.jpg'
img = imread(path)
# -1 表示人脸检测的判定阈值
# scores 为每个检测结果的得分，idx 为人脸检测器的类型
dets, scores, idx = detector.run(img, 1, -1)
for i, d in enumerate(dets):
    print('%d: score %f, face_type %f' % (i, scores[i], idx[i]))
win.clear_overlay()
win.set_image(img)
win.add_overlay(dets)
dlib.hit_enter_to_continue()
```

执行代码后会检测图片 111.jpg 中的人脸，执行效果如图 4-1 所示。

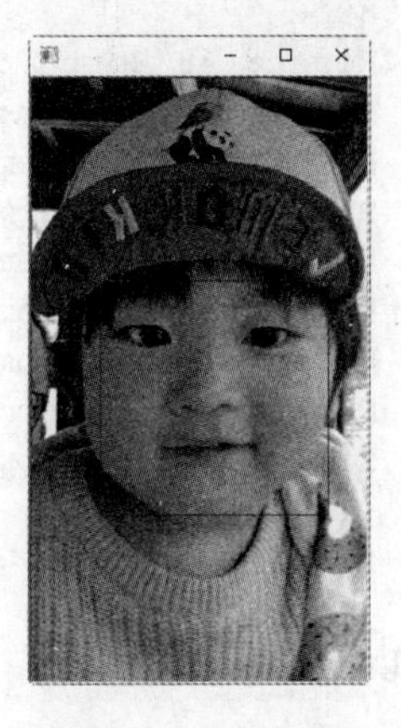

图 4-1　执行效果

在 dlib 中，通过使用内置函数 find_candidate_object_locations()可识别出指定的图片，并把可能存在的目标对象找出来。函数 find_candidate_object_locations()可以快速找到候选目标的区域，然后可以使用这些区域进行后续操作。请看下面的实例文件

find_candidate_object_locations.py，演示了使用函数 find_candidate_object_locations()识别图片 111.jpg 的过程。

源码路径：daima\4\find_candidate_object_locations.py

```
import dlib

image_file = '111.jpg'
img = dlib.load_rgb_image(image_file)

rects = []
dlib.find_candidate_object_locations(img, rects, min_size=500)

print("number of rectangles found {}".format(len(rects)))
for k, d in enumerate(rects):
    print("Detection {}: Left: {} Top: {} Right: {} Bottom: {}".format(
        k, d.left(), d.top(), d.right(), d.bottom()))
```

在上述代码中，函数 load_rgb_image()接收一个文件名称，然后返回 NumPy 数组对象，这个返回值作为函数 find_candidate_object_locations()的输入。函数 find_candidate_object_locations()的第二个参数 rects 为列表，用于保存找到候选对象所在的区域。第三个参数 min_size 表示找到的区域大小不应该小于指定的像素值。

通过上述代码，候选对象的位置将保存到矩形中。执行代码后会输出：

```
number of rectangles found 889
Detection 0: Left: 0 Top: 0 Right: 68 Bottom: 34
Detection 1: Left: 0 Top: 0 Right: 94 Bottom: 34
Detection 2: Left: 0 Top: 0 Right: 117 Bottom: 36
Detection 3: Left: 0 Top: 0 Right: 119 Bottom: 34
Detection 4: Left: 0 Top: 0 Right: 120 Bottom: 45
Detection 5: Left: 0 Top: 0 Right: 120 Bottom: 103
Detection 6: Left: 0 Top: 0 Right: 141 Bottom: 100
Detection 7: Left: 0 Top: 0 Right: 141 Bottom: 103
Detection 8: Left: 0 Top: 0 Right: 141 Bottom: 137
……后面省略
```

4.2.2 使用命令行的人脸识别

在上一个实例中，在代码中设置了要识别的图片文件是 111.jpg。我们也可以使用命令行的格式识别指定的图片，请看下面的实例文件 face_detector.py，演示了通过命令行对指定图片实现人脸检测的过程。本实例的人脸检测器采用了经典的方向梯度直方图(HOG)特征，结合线性分类器、图像金字塔和滑动窗口检测方案进行识别。这种类型的目标探测器相当普遍，能够检测除了人脸外的许多类型的半刚性物体。文件 face_detector.py 的具体实现代码如下。

源码路径：daima\4\face_detector.py

```
import sys
import dlib
detector = dlib.get_frontal_face_detector()
win = dlib.image_window()

for f in sys.argv[1:]:
    print("Processing file: {}".format(f))
    img = dlib.load_rgb_image(f)
    #第二个参数中的 1 表示应该将图像向上采样 1 次，这将能够发现更多的面孔
    dets = detector(img, 1)
    print("Number of faces detected: {}".format(len(dets)))
    for i, d in enumerate(dets):
        print("Detection {}: Left: {} Top: {} Right: {} Bottom: {}".format(
            i, d.left(), d.top(), d.right(), d.bottom()))

    win.clear_overlay()
    win.set_image(img)
    win.add_overlay(dets)
    dlib.hit_enter_to_continue()

if (len(sys.argv[1:]) > 0):
    img = dlib.load_rgb_image(sys.argv[1])
    dets, scores, idx = detector.run(img, 1, -1)
    for i, d in enumerate(dets):
        print("Detection {}, score: {}, face_type:{}".format(
            d, scores[i], idx[i]))
```

在上述代码的 if 语句中，可以让探测器告诉你每次检测的分数。分数越大，表示越有信心探测到。在函数 detector.run()中，第三个参数是对检测阈值的可选调整，其中负值将返回更多的检测，而正值将返回更少的检测。另外，变量 idx 用于设置匹配哪些人脸子探测器，可用于识别不同方向的人脸。输入下面的命令可以检测图片 222.jpg 中的人脸：

```
python face_detector.py 222.jpg
```

运行上述命令行命令后显示下面的检测信息，并显示检测结果，如图 4-2 所示。

```
Processing file: 222.jpg
Number of faces detected: 1
Detection 0: Left: 66 Top: 32 Right: 118 Bottom: 84
Hit enter to continue
Detection [(66, 32) (118, 84)], score: 2.6202281155122633, face_type:0
Detection [(129, 118) (181, 170)], score: -0.6346316183738288, face_type:1
```

图 4-2　检测结果

4.2.3　检测人脸关键点

使用训练好的模型 shape_predictor_68_face_landmarks.dat(人脸识别 68 个特征点检测数据库)检测人脸关键点，在检测出人脸的同时，可检测出人脸上的 68 个关键点。开发者可以在 dlib 的官方网站下载模型文件 shape_predictor_68_face_landmarks.dat，下载网址是 http://dlib.net/files/，如图 4-3 所示。

```
dlib-19.5.tar.bz2
dlib-19.5.zip
dlib-19.6.tar.bz2
dlib-19.6.zip
dlib-19.7.tar.bz2
dlib-19.7.zip
dlib-19.8.tar.bz2
dlib-19.8.zip
dlib-19.9.tar.bz2
dlib-19.9.zip
dlib_documentation-18.16.chm
dlib_documentation-18.17.chm
dlib_documentation-18.18.chm
dlib_face_recognition_resnet_model_v1.dat.bz2
dlib_face_recognition_resnet_model_v1_lfw_test_scripts.tar.bz2
dlib_kitti_submission_mmodCNN_basic7convModel.tar.bz2
imagenet2015_validation_images.txt.bz2
instance_segmentation_voc2012net.dnn
instance_segmentation_voc2012net_v2.dnn
mmod_dog_hipsterizer.dat.bz2
mmod_front_and_rear_end_vehicle_detector.dat.bz2
mmod_human_face_detector.dat.bz2
mmod_rear_end_vehicle_detector.dat.bz2
resnet34_1000_imagenet_classifier.dnn.bz2
resnet50_1000_imagenet_classifier.dnn.bz2
semantic_segmentation_voc2012net.dnn
```

图 4-3　dlib 官方网站的模型文件

在 dlib 中，使用内置函数 shape_predictor()可实现预测器功能，在图片中标记人脸关键点。函数 shape_predictor()的具体格式如下：

```
dlib.shape_predictor('data/data_dlib/shape_predictor_68_face_landmarks.dat')
```

参数说明如下。

- 参数 data/data_dlib/shape_predictor_68_face_landmarks.dat：68 个关键点模型地址。
- 返回值：人脸关键点预测器。

编写实例文件 second.py，功能是使用模型 shape_predictor_68_face_landmarks.dat 检测人脸关键点。

源码路径：daima\4\second.py

(1) 首先使用 import 语句导入 dlib 模块，代码如下：

```
import dlib
from imageio import imread
import glob
```

(2) 分别准备好人脸检测器、关键点检测模型、显示窗口和要检测的图片路径，代码如下：

```
detector = dlib.get_frontal_face_detector()
predictor_path = 'shape_predictor_68_face_landmarks.dat'
predictor = dlib.shape_predictor(predictor_path)
win = dlib.image_window()
path ='111.jpg'

img = imread(path)
win.clear_overlay()
win.set_image(img)
```

(3) 检测图片 111.jpg 中的人脸关键点，代码如下：

```
detector = dlib.get_frontal_face_detector()
predictor_path = 'shape_predictor_68_face_landmarks.dat'
predictor = dlib.shape_predictor(predictor_path)
win = dlib.image_window()
path ='111.jpg'

img = imread(path)
win.clear_overlay()
win.set_image(img)

# 1 表示将图片放大一倍，便于检测到更多人脸
dets = detector(img, 1)
```

```
print('检测到了 %d 个人脸' % len(dets))
for i, d in enumerate(dets):
        print('- %d: Left %d Top %d Right %d Bottom %d' % (i, d.left(), d.top(),
              d.right(), d.bottom()))
        shape = predictor(img, d)
        # 第 0 个点和第 1 个点的坐标
        print('Part 0: {}, Part 1: {}'.format(shape.part(0), shape.part(1)))
        win.add_overlay(shape)

win.add_overlay(dets)
dlib.hit_enter_to_continue()
```

在上述代码中，使用函数 shape_predictor()在检测出人脸基础上找到人脸的 68 个特征点。执行代码后会检测图片 111.jpg 中的人脸关键点，执行效果如图 4-4 所示，并输出下面的内容：

```
检测到了 1 个人脸
- 0: Left 56 Top 160 Right 242 Bottom 345
Part 0: (35, 200), Part 1: (37, 229)
Hit enter to continue
```

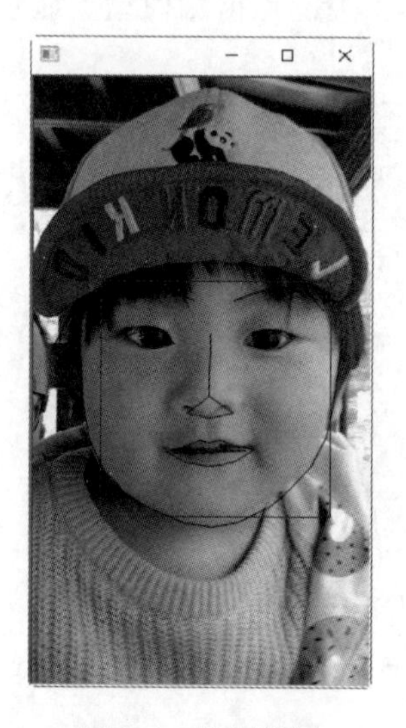

图 4-4　执行效果

4.2.4　基于 CNN 的人脸检测器

基于机器学习 CNN 方法来检测人脸比前面介绍的方法效率要慢很多，在下面的实例文件 cnn_face_detector.py 中，演示了使用 dlib 运行基于 CNN 的人脸检测器。本实例加载一个经过预训练的模型 mmod_human_face_detector.dat，并使用它在图像中查找人脸。文件 cnn_face_detector.py 的具体实现代码如下。

源码路径：daima\4\cnn_face_detector.py

```
import sys
import dlib
if len(sys.argv) < 3:
    print(
        "Call this program like this:\n"
        "   ./cnn_face_detector.py mmod_human_face_detector.dat ../
                              examples/faces/*.jpg\n"
        "You can get the mmod_human_face_detector.dat file from:\n"
        "    http://dlib.net/files/mmod_human_face_detector.dat.bz2")
    exit()

cnn_face_detector = dlib.cnn_face_detection_model_v1(sys.argv[1])
win = dlib.image_window()

for f in sys.argv[2:]:
    print("Processing file: {}".format(f))
    img = dlib.load_rgb_image(f)
    dets = cnn_face_detector(img, 1)
    print("Number of faces detected: {}".format(len(dets)))
    for i, d in enumerate(dets):
        print("Detection {}: Left: {} Top: {} Right: {} Bottom: {} Confidence:
              {}".format(i, d.rect.left(), d.rect.top(), d.rect.right(),
              d.rect.bottom(), d.confidence))

    rects = dlib.rectangles()
    rects.extend([d.rect for d in dets])

    win.clear_overlay()
    win.set_image(img)
    win.add_overlay(rects)
    dlib.hit_enter_to_continue()
```

上述检测器会返回一个 mmod_rectangles 矩阵对象，此对象包含 mmod_rectangles 对象的列表，可以通过迭代来访问 mmod_rectangles 对象。mmod_rectangles 对象有两个成员变量，一个 dlib.rectangle 对象和一个匹配度得分。输入下面的命令运行本实例程序：

```
python cnn_face_detector.py mmod_human_face_detector.dat 111.jpg
```

通过运行上述命令行命令，可以基于 CNN 对图片 111.jpg 实现人脸检测，输出下面的检测结果，执行效果如图 4-5 所示。

```
Processing file: 111.jpg
Number of faces detected: 1
Detection 0: Left: 39 Top: 120 Right: 243 Bottom: 324 Confidence:
1.0611059665679932
```

图 4-5　执行效果

4.2.5　在摄像头中识别人脸

在现实应用中，经常需要识别摄像头中的人脸。在下面的实例文件 opencv_webcam_face_detection.py 中，使用 OpenCV 和 dlib 在网络摄像头中查找正面人脸。这也意味着 dlib 的 rgb 图像可以与 OpenCV 一起使用，只需交换红色和蓝色通道即可。

源码路径：daima\4\opencv_webcam_face_detection.py

```
detector = dlib.get_frontal_face_detector()
cam = cv2.VideoCapture(0)
color_green = (0,255,0)
line_width = 3
while True:
    ret_val, img = cam.read()
    rgb_image = cv2.cvtColor(img, cv2.COLOR_BGR2RGB)
    dets = detector(rgb_image)
    for det in dets:
        cv2.rectangle(img,(det.left(), det.top()), (det.right(), det.bottom()),
                    color_green, line_width)
    cv2.imshow('my webcam', img)
    if cv2.waitKey(1) == 27:
        break  # esc to quit
cv2.destroyAllWindows()
```

执行效果如图 4-6 所示。

图 4-6 执行效果

4.2.6 人脸识别验证

在上一个实例的基础上将人脸提取为特征向量，从而可以对特征向量进行比对来实现人脸的验证。请看下面的实例，功能是验证两张照片是否是同一个人。本实例采用的是对比欧式距离的方法，编写实例文件 shibie.py，使用训练好的 ResNet 人脸识别模型文件 dlib_face_recognition_resnet_model_v1.dat 进行识别，代码如下：

源码路径：daima\4\shiebie.py

```
import dlib
from imageio import imread
import numpy as np

detector = dlib.get_frontal_face_detector()
predictor_path = 'shape_predictor_68_face_landmarks.dat'
predictor = dlib.shape_predictor(predictor_path)
face_rec_model_path = 'dlib_face_recognition_resnet_model_v1.dat'
facerec = dlib.face_recognition_model_v1(face_rec_model_path)


def get_feature(path):
    img = imread(path)
    dets = detector(img)
    print('检测到了 %d 个人脸' % len(dets))
    # 这里假设每张图只有一个人脸
    shape = predictor(img, dets[0])
    face_vector = facerec.compute_face_descriptor(img, shape)
    return(face_vector)

def distance(a,b):
    a,b = np.array(a), np.array(b)
```

```
        sub = np.sum((a-b)**2)
        add = (np.sum(a**2)+np.sum(b**2))/2.
        return sub/add

    path_lists1 = ["f1.jpg","f2.jpg"]
    path_lists2 = ["毛毛照片.jpg","毛毛测试.jpg"]

    feature_lists1 = [get_feature(path) for path in path_lists1]
    feature_lists2 = [get_feature(path) for path in path_lists2]

    print("feature 1 shape",feature_lists1[0].shape)

    out1 = distance(feature_lists1[0],feature_lists1[1])
    out2 = distance(feature_lists2[0],feature_lists2[1])

    print("diff distance is",out1)
    print("same distance is",out2)

    def classifier(a,b,t = 0.09):
        if(distance(a,b)<=t):
            ret = True
        else :
            ret = False
        return(ret)

    print("f1 is 毛毛",classifier(feature_lists1[0],feature_lists2[1]))
    print("f2 is 毛毛",classifier(feature_lists1[1],feature_lists2[1]))
    print("毛毛照片.jpg is 毛毛.jpg",classifier(feature_lists2[0],feature_lists2[1]))
```

通过上述代码，每张人脸都被提取为 128 维的向量，我们可以将其理解为 128 维的坐标(xyz 是三维，128 维就是由 128 个轴组成)。然后，计算两个特征的距离，设定好合适的阈值，如果小于这个阈值则识别为同一个人。执行后会输出：

```
检测到了 1 个人脸
检测到了 1 个人脸
检测到了 1 个人脸
检测到了 1 个人脸
feature 1 shape (128, 1)
diff distance is 0.25476771591192765
same distance is 0.06
f1 is 毛毛 False
f2 is 毛毛 False
毛毛照片.jpg is 毛毛.jpg True
```

通过上述执行效果可以看出，不同的距离为 0.25476771591192765，同一个人为 0.06，我们可以先将阈值设置为其间的一个值。在上述代码中设置为 0.09，这个阈值也是需要大量数据来计算的，选择的准则为使错误识别为最低。在将阈值设置为 0.09 后，使用函数 classifier(a,b,t = 0.09)测试能否区分出不同的人。通过上述实例的执行效果可以看出，本实例代码已基本满足对人脸区分的功能，如果想要商业化应用，则需要继续调优阈值与代码，调优的准则就是选择合适的阈值使错误识别率为最低。

4.2.7 全局优化

所有机器学习开发者都会遇到同样一个问题：有一些想要使用的机器学习算法，但其中填满了超参数——这些数字包括权重衰减率、高斯核函数宽度，等等。算法本身并不会设置它们，开发者必须自己决定它们的数值。如果调的参数不够好，那么算法是不会工作的。在调参时，绝大多数人只会凭经验进行猜测。很显然这不是一个好现象，我们需要更合理的方法。

dlib 作为一个开源的 C++ 机器学习算法工具包，被广泛用于工业界和学术界，覆盖机器人、嵌入式设备、手机和大型高性能计算设备等环境。dlib 从 v19.8 版本开始，为开发者引入了自动调优超参数的 LIPO 算法。dlib 中的调优方法最大优势是简单，非常易于进行超参数优化的工作。

请看下面的实例文件 global_optimization.py，功能是使用 dlib 内置的全局优化函数 dlib.find_min_global()查找自定义函数 holder_table()的输入，从而使函数 holder_table()的输出最小。dlib 的全局优化是一个非常有用的工具，便于应用机器学习函数的超参搜索功能。本实例只是演示了如何调用全局优化方法，使用一个通用的全局优化测试函数 find_min_global()调用定义函数 holder_table()。文件 global_optimization.py 的具体实现代码如下。

源码路径：daima\4\global_optimization.py

```
import dlib
from math import sin,cos,pi,exp,sqrt

# 这是针对优化问题的标准测试函数，它有一组局部最小值，设置全局最小值保持器
holder_table()==-19.2085025679.
def holder_table(x0,x1):
    return -abs(sin(x0)*cos(x1)*exp(abs(1-sqrt(x0*x0+x1*x1)/pi)))

#找到 holder\u table()的最佳输入
#后面的 print 语句显示 find_min_global()以提高精度查找最佳设置
x,y = dlib.find_min_global(holder_table,
                           [-10,-10],  #x0 和 x1 的下界约束
```

```
                    [10,10],     #x0和x1的上界约束
                    80)         #find_min_global()调用holder_table()的次数

print("optimal inputs: {}".format(x));
print("optimal output: {}".format(y));
```

执行后会输出：

```
optimal inputs: [8.057092112728306, 9.67043010020526]
optimal output: -19.208113390882694
```

另外，还可以使用 dlib 的内置函数 max_cost_assignment()计算分配任务的最大价值方案。请考虑下面的场景：

需要将 N 个人分配给 N 个工作，每个人在每一份工作上都会给公司赚一定的钱。但是因为每个人的技能不同，所以他们在某些工作上做得更好，在另一些工作上做得不好。请想办法找到最好的方法来为这些人分配工作，最大限度地提高公司作为一个整体的盈利。

在这个问题中，假设有 3 个人和 3 份工作。用一个矩阵来表示每个人在每项工作中的收入，每行对应一个人，每列对应一个工作。比如员工 0 在工作 0 赚 1 美元，在工作 1 赚 2 美元，在工作 2 赚 6 美元。

请看下面的实例文件 max_cost_assignment.py，展示了调用函数 max_cost_assignment()解决上述最优线性分配求解器问题的过程。本算法是匈牙利算法的一个实现，运行速度非常快，时间复杂度为 $O(N^3)$。实例文件 max_cost_assignment.py 的具体实现代码如下。

源码路径：daima\4\max_cost_assignment.py

```
import dlib

#工作矩阵
cost = dlib.matrix([[1, 2, 6],
                    [5, 3, 6],
                    [4, 5, 0]])

#调用函数dlib.max_cost_assignment()找出最佳的工作分配
assignment = dlib.max_cost_assignment(cost)

# 打印最佳分配：[2, 0, 1]
print("Optimal assignments: {}".format(assignment))
print("Optimal cost: {}".format(dlib.assignment_cost(cost, assignment)))
```

执行后会输出：

```
Optimal assignments: [2, 0, 1]
Optimal cost: 16.0
```

通过上述执行结果可知，输出的最佳分配是[2，0，1]，这表示应该将成本矩阵第一行的人员分配给作业 2，将中间行的人员分配给作业 0，将底部行的人员分配给作业 1。输出显示的最佳成本是 16.0，这是正确的，因为我们的最佳分配是 6+5+5。

4.2.8 人脸聚类

如果在一个图片中有大量的人脸，我们可以基于人脸识别标准进行聚类操作，把距离较近的人脸聚为一类，即识别为有可能是同一个人。请看下面的实例，功能是将某个目录下的所有照片中被认为是同一个人的人脸提取出来。实例文件 juface.py 的具体实现代码如下。

源码路径：daima\4\juface.py

(1) 下载模型文件 shape_predictor_68_face_landmarks.dat 和 dlib_face_recognition_resnet_model_v1.dat，设置要处理的图片目录为 paths。代码如下：

```
detector = dlib.get_frontal_face_detector()
predictor_path = 'shape_predictor_68_face_landmarks.dat'
predictor = dlib.shape_predictor(predictor_path)
face_rec_model_path = 'dlib_face_recognition_resnet_model_v1.dat'
facerec = dlib.face_recognition_model_v1(face_rec_model_path)
paths = glob.glob('faces/*.jpg')
```

(2) 获取所有图片的关键点检测结果和向量表示，代码如下：

```
vectors = []
images = []
for path in paths:
    img = imread(path)
    dets = detector(img, 1)
    for i, d in enumerate(dets):
        shape = predictor(img, d)
        face_vector = facerec.compute_face_descriptor(img, shape)
        vectors.append(face_vector)
        images.append((img, shape))
```

(3) 以 0.5 为阈值进行聚类，并找出人脸数量最多的类。代码如下：

```
labels = dlib.chinese_whispers_clustering(vectors, 0.5)
num_classes = len(set(labels))
print('共聚为 %d 类' % num_classes)
biggest_class = Counter(labels).most_common(1)
print(biggest_class)
```

(4) 将最大类中包含的人脸保存下来，代码如下：

```
output_dir = 'most_common'
if not os.path.exists(output_dir):
    os.mkdir(output_dir)
face_id = 1
for i in range(len(images)):
    if labels[i] == biggest_class[0][0]:
        img, shape = images[i]
        dlib.save_face_chip(img, shape, output_dir + '/face_%d' % face_id,
                            size=150, padding=0.25)
        face_id += 1
```

执行代码后会聚类处理某个目录下的所有照片，将照片中的被认为是同一个人的人脸提取出来，然后保存到 most_common 目录，如图 4-7 所示。

图 4-7　被认为是同一个人的人脸

4.2.9　抖动采样和增强

请看下面的实例文件 face_jitter.py，展示了使用 dlib 人脸识别模型训练数据，对指定图像中的人脸进行抖动采样和增强处理的过程。本实例可以接收输入指定的图像并干扰颜色，同时应用随机平移、旋转和缩放操作。实例文件 face_jitter.py 的具体实现代码如下。

源码路径：daima\4\face_jitter.py

```
import sys
import dlib
def show_jittered_images(window, jittered_images):
    '''
        逐一显示指定的抖动图像
    '''
    for img in jittered_images:
        window.set_image(img)
        dlib.hit_enter_to_continue()

if len(sys.argv) != 2:
    print(
        "Call this program like this:\n"
        "   ./face_jitter.py shape_predictor_5_face_landmarks.dat\n"
```

```
        "You can download a trained facial shape predictor from:\n"
        "    http://dlib.net/files/shape_predictor_5_face_landmarks.dat.bz2\n")
    exit()

predictor_path = sys.argv[1]
face_file_path = "111.jpg"

#加载需要的所有模型：一个检测器查找人脸，一个形状预测器查找人脸标志，这样就可以精确地定位人脸
detector = dlib.get_frontal_face_detector()
sp = dlib.shape_predictor(predictor_path)

#使用 dlib 加载图像
img = dlib.load_rgb_image(face_file_path)

#让探测器找到每个面的边界框
dets = detector(img)
num_faces = len(dets)
# 找到 5 个面部标志，我们需要做的路线
faces = dlib.full_object_detections()
for detection in dets:
    faces.append(sp(img, detection))

# 获取对齐的人脸图像并显示出来
image = dlib.get_face_chip(img, faces[0], size=320)
window = dlib.image_window()
window.set_image(image)
dlib.hit_enter_to_continue()

#显示 5 个抖动的图像而不增加数据
jittered_images = dlib.jitter_image(image, num_jitters=5)
show_jittered_images(window, jittered_images)

#显示 5 个抖动的图像和数据增强
jittered_images = dlib.jitter_image(image, num_jitters=5,
disturb_colors=True)
show_jittered_images(window, jittered_images)
```

通过如下命令运行本实例程序，执行后可以将图片 111.jpg 采样，执行效果如图 4-8 所示。

```
python face_jitter.py shape_predictor_5_face_landmarks.dat
```

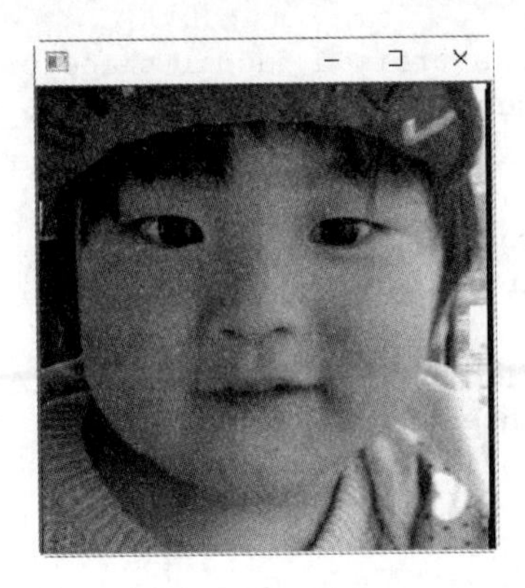

图 4-8　执行效果

4.2.10　人脸和姿势采集

请看下面的实例文件 face_landmark_detection.py，功能是在指定的图像中找到正面人脸并估计他们的姿势。在文件 shape_predictor_68_face_landmarks.dat 中保存了人脸 68 点特征检测器数据集，这些数据集是脸上的点，比如嘴角、眉毛、眼睛等。

源码路径：daima\4\face_landmark_detection.py

```
if len(sys.argv) != 3:
    print(
        "Give the path to the trained shape predictor model as the first "
        "argument and then the directory containing the facial images.\n"
        "For example, if you are in the python_examples folder then "
        "execute this program by running:\n"
        "    ./face_landmark_detection.py shape_predictor_68_face_
                          landmarks.dat ../examples/faces\n"
        "You can download a trained facial shape predictor from:\n"
        "    http://dlib.net/files/shape_predictor_68_face_landmarks.dat.bz2")
    exit()

predictor_path = sys.argv[1]
faces_folder_path = sys.argv[2]

detector = dlib.get_frontal_face_detector()
predictor = dlib.shape_predictor(predictor_path)
win = dlib.image_window()

for f in glob.glob(os.path.join(faces_folder_path, "*.jpg")):
    print("Processing file: {}".format(f))
    img = dlib.load_rgb_image(f)
```

```
    win.clear_overlay()
    win.set_image(img)

    # 让探测器找到每个面的边界框。第二个参数中的 1 表示应该将图像向上采样 1 次。这将使范围
    # 变得更大，让我们能够发现更多的面孔。
    dets = detector(img, 1)
    print("Number of faces detected: {}".format(len(dets)))
    for k, d in enumerate(dets):
        print("Detection {}: Left: {} Top: {} Right: {} Bottom: {}".format(
            k, d.left(), d.top(), d.right(), d.bottom()))
        #在方框 d 中获取面部的 landmarks/parts
        shape = predictor(img, d)
        print("Part 0: {}, Part 1: {} ...".format(shape.part(0), shape.part(1)))
        #画出面部标记
        win.add_overlay(shape)

    win.add_overlay(dets)
    dlib.hit_enter_to_continue()
```

本实例使用的人脸检测器是利用了经典的方向梯度直方图(HOG)特征，结合线性分类器、图像金字塔和滑动窗口检测方案。假设在 faces 目录中保存了多个图片文件，通过如下命令可以提取照片中的人脸和姿势，执行效果如图 4-9 所示。按回车键后，会继续识别 faces 目录中的下一张照片。

```
python face_landmark_detection.py shape_predictor_68_face_landmarks.dat faces
```

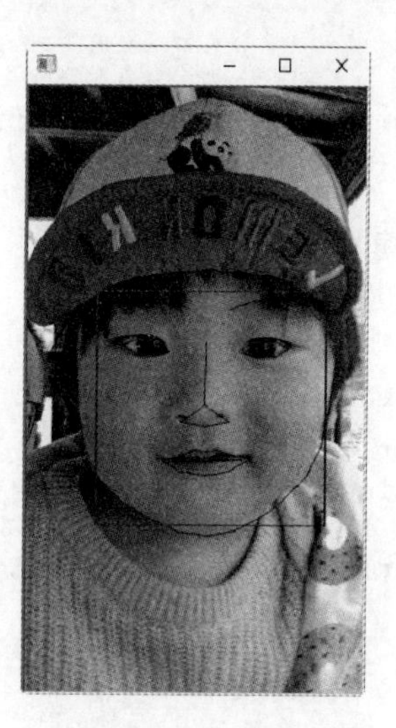

图 4-9　执行效果

4.2.11 物体追踪

物体追踪是指对于视频文件，在第一帧指定一个矩形区域，对于后续帧自动追踪和更新区域的位置。使用 Python 库 dlib 中的 correlation 跟踪器，可以实时跟踪检测视频中某移动对象的位置。在使用 correlation 跟踪器时，需要将当前视频帧中要跟踪的对象的边界框指定给“相关性跟踪程序”，然后在随后的帧中识别这个对象的位置。假设在一张桌子上放了一个果汁盒和其他物品，然后用移动的相机拍摄这个桌子上的物品。请看下面的实例文件 wuzhui.py，我们使用 dlib 追踪视频中的果汁盒。

源码路径：daima\4\wuzhui.py

```
# 视频帧的路径
video_folder = os.path.join("video_frames")

#创建相关跟踪器，初始化对象后才能使用
tracker = dlib.correlation_tracker()
#准备好追踪器和图片
win = dlib.image_window()
#将从磁盘加载帧时跟踪它们
for k, f in enumerate(sorted(glob.glob(os.path.join(video_folder, "*.jpg")))):
    print("Processing Frame {}".format(k))
    img = dlib.load_rgb_image(f)

    # 需要在第一帧初始化跟踪器
    if k == 0:
        # 开始追踪果汁盒，如果看第一帧，会看到果汁盒所在位置的边界框是(74, 67, 112, 153)
        tracker.start_track(img, dlib.rectangle(74, 67, 112, 153))
    else:
        # 否则就从上一帧开始跟踪
        tracker.update(img)

    win.clear_overlay()
    win.set_image(img)
    win.add_overlay(tracker.get_position())
    dlib.hit_enter_to_continue()
```

执行代码后可以追踪 video_frames 目录中的视频帧，标记出每一帧图像中果汁盒的位置，执行效果如图 4-10 所示。

图 4-10 执行效果

4.3 SVM 分类算法

SVM 是支持向量机(Support Vector Machine)的缩写，是一类按监督学习(Supervised Learning)方式对数据进行二元分类的广义线性分类器(Generalized Linear Classifier)，其决策边界是对学习样本求解的最大边距超平面(Maximum-margin Hyperplane)。SVM 使用铰链损失函数(Hinge Loss)计算经验风险，并在求解系统中加入了正则化项以优化结构风险，是一个具有稀疏性和稳健性的分类器。SVM 可以通过核方法(Kernel Method)进行非线性分类，是常见的核学习(Kernel Learning)方法之一。

扫码观看本节视频讲解

4.3.1 二进制 SVM 分类器

请看下面的实例文件 svm_binary_classifier.py，使用 dlib 的内置库实现二进制 SVM 分类器。在本实例中创建了一个简单的测试数据集，展示了实现简易 SVM 分类器的方法。文件 svm_binary_classifier.py 的具体实现流程如下。

源码路径：daima\4\svm_binary_classifier.py

(1) 导入需要的库，创建两个训练数据集。在现实应用中，通常会使用更大的训练数据集，但是就本实例来说，两个训练例子就已经足够了。对于二进制分类器而言，y 标签应该都是+1 或-1。代码如下：

```
import dlib
try:
```

```
   import cPickle as pickle
except ImportError:
   import pickle

x = dlib.vectors()
y = dlib.array()

# x.append(dlib.vector([1, 2, 3, -1, -2, -3]))
y.append(+1)

x.append(dlib.vector([-1, -2, -3, 1, 2, 3]))
y.append(-1)
```

(2) 制作一个训练对象，此对象负责将训练数据集转换为预测模型。本实例是一个使用线性核的支持向量机训练器，如果要使用 RBF 内核或直方图相交内核，可以将其更改为以下之一的代码行：

- svm = dlib.svm_c_trainer_histogram_intersection()
- svm = dlib.svm_c_trainer_radial_basis()

代码如下：

```
svm = dlib.svm_c_trainer_linear()
svm.be_verbose()
svm.set_c(10)
```

(3) 开始训练模型，返回值是能够进行预测的训练模型，然后用我们的数据运行模型并查看结果。代码如下：

```
classifier = svm.train(x, y)

#查看结果
print("prediction for first sample:  {}".format(classifier(x[0])))
print("prediction for second sample: {}".format(classifier(x[1])))
```

(4) 也可以像任何其他 Python 对象一样，使用 Python 内置库 pickle 序列化分类器模型对象。代码如下：

```
with open('saved_model.pickle', 'wb') as handle:
   pickle.dump(classifier, handle, 2)
```

执行代码后会输出：

```
objective:    0.0178571
objective gap: 0
risk:         0
risk gap:     0
```

```
num planes:    3
iter:        1

prediction for first sample:  1.0
prediction for second sample: -1.0
```

4.3.2 Ranking SVM 算法

Learning To Rank(简称 LTR)是用机器学习的思想来解决排序问题。LTR 有 3 种主要的方法：PointWise，PairWise 和 ListWise。Ranking SVM 算法是 PairWise 方法的一种，由 R. Herbrich 等人在 2000 年提出，T. Joachims 介绍了一种基于用户 Clickthrough 数据使用 Ranking SVM 来进行排序的方法(SIGKDD, 2002)。

在 dlib 的 C++库中内置了 SVM-Rank 工具，这是一个学习排列对象的有用工具。例如，可以让 SVM-Rank 学习根据用户的搜索引擎结果对网页进行排名，其思想是将最相关的页面排名高于不相关的页面。

在下面的实例文件 svm_rank.py 中，我们将创建一个简单的测试数据集，并使用机器学习方法来学习一个函数。函数的目的是给“相关”对象比“非相关”对象更高的分数。其思想是使用此分数对对象进行排序，以便最相关的对象位于排名列表的顶部。文件 svm_rank.py 的具体实现流程如下所示。

源码路径：daima\4\svm_rank.py

(1) 首先准备测试数据。为了简单起见，假设需要对二维向量进行排序，并且在第一维中具有正值的向量的排序高于其他向量。因此我们要做的是制作相关(即高排名)和非相关(即低排名)向量的示例，并将它们存储到排名对象中。代码如下：

```
data = dlib.ranking_pair()
# 添加两个示例。在实际应用中，可能需要大量相关和非相关向量的示例
data.relevant.append(dlib.vector([1, 0]))
data.nonrelevant.append(dlib.vector([0, 1]))
```

(2) 现在我们有了一些数据，接下来可以使用机器学习方法来学习一个函数，该函数的功能是给相关向量高分，给非相关向量低分。代码如下：

```
trainer = dlib.svm_rank_trainer()
# trainer 对象具有一些控制其行为的参数
#由于这是 SVM-Rank 算法，所以需要使用参数 c 来控制，在尝试精确拟合训练数据或选择一个“更简单”的
#解决方案之间的权衡
trainer.c = 10
```

(3) 使用函数 train()开始训练上面的数据，如果在向量上调用 rank 将输出一个排名分数，相关向量的排名得分应当大于非相关向量的得分。代码如下：

```
rank = trainer.train(data)

print("相关向量排名得分:    {}".format(
   rank(data.relevant[0])))
print("非相关向量的排名得分: {}".format(
   rank(data.nonrelevant[0])))
```

(4) 如果想要一个排名精度的整体度量，可以通过调用函数 test_ranking_function()来计算排序精度和平均精度值。在这种情况下，排序精度告诉我们非相关向量排在相关向量前面的频率。在本实例中，函数 test_ranking_function()为这两个度量返回 1，表示使用 rank 函数输出一个完美的排名。代码如下：

```
print(dlib.test_ranking_function(rank, data))

#排名得分是通过获取学习的权重向量和数据向量之间的点积来计算的。如果想查看学习的权重向量，
#可以这样显示:
print("Weights: {}".format(rank.weights))#在这种情况下的权重
```

(5) 在上面的示例中，我们的数据只包含两组对象：相关集和非相关集。训练正试图找到一个排序函数，该函数使每个相关向量的得分高于每个非相关向量的得分。但是在现实应用中，有时候想要做的事情远比这要复杂一些。例如，在某用户浏览 Web 页面的排名应用中，我们必须根据用户查询的页面进行排名。在这种情况下，每个查询都有自己的一组相关和非相关文档，与一个查询相关的内容很可能与另一个查询无关。所以此时我们没有一组全局相关网页和另一组非相关网页的数据。要处理这样的情况，我们可以简单地给训练提供多个 ranking_u_pair 排序对实例。因此，每个排序对表示特定查询的“相关/非相关”集。例如下面的排序对实例，为了简单起见，我们重用了上面的数据来进行 4 个相同的“查询”。

```
queries = dlib.ranking_pairs()
queries.append(data)
queries.append(data)
queries.append(data)
queries.append(data)

#像以前一样训练
rank = trainer.train(queries)
```

(6) 现在我们有了多个 ranking_u_pair 排序对实例，可以使用函数 cross_validate_

ranking_trainer()将查询拆分为多个折叠来执行交叉验证。也就是说，它可以让训练对一个子集进行训练，并对其他实例进行测试。本实例将通过 4 个不同的子集来实现这一点，并根据保留的数据返回总体排名精度。代码如下：

```
# 与 test_ranking_function()一样，同时报告排序精度和平均精度
print("Cross 交叉验证结果: {}".format(
    dlib.cross_validate_ranking_trainer(trainer, queries, 4)))
```

(7) 除了在上面使用过的密集向量之外，排名工具还支持使用稀疏向量。因此通过下面的代码，可以跟上面示例程序的第一部分那样使用稀疏向量：

```
data = dlib.sparse_ranking_pair()
samp = dlib.sparse_vector()
```

(8) 使 samp 表示与 dlib.vector([1, 0])相同的向量。

代码如下：

```
samp.append(dlib.pair(0, 1))
data.relevant.append(samp)
```

在 dlib 中，稀疏向量只是由成对对象组成的数组，每对存储一个索引和一个值。此外，支持向量机排序工具需要对稀疏向量进行排序，并具有唯一的索引。这意味着索引是按递增顺序展示的，索引值不会出现一次以上。在日常应用中，可以使用 dlib.make_sparse_vector()使稀疏向量对象正确排序并包含唯一的索引。

(9) 我们可以让 samp 表示与 dlib.vector 相同的向量([0, 1])，最后使用函数 train()开始训练上面的数据。代码如下：

```
samp.clear()
samp.append(dlib.pair(1, 1))
data.nonrelevant.append(samp)

trainer = dlib.svm_rank_trainer_sparse()
rank = trainer.train(data)
print("相关向量的排名分数:     {}".format(
    rank(data.relevant[0])))
print("非相关向量的排名分数: {}".format(
    rank(data.nonrelevant[0])))
```

执行代码后会输出：

```
相关向量排名得分:     0.5
非相关向量的排名得分: -0.5
ranking_accuracy: 1  mean_ap: 1
Weights: 0.5
```

```
-0.5
Cross 交叉验证结果: ranking_accuracy: 1  mean_ap: 1
相关向量的排名分数:     0.5
非相关向量的排名分数: -0.5
```

4.3.3 Struct SVM 多分类器

通过使用 dlib 内置的 SVM-Rank 工具，可以实现 Struct SVM 多分类器。请看下面的实例文件 svm_struct.py，演示了使用 Struct SVM 多分类器的知识。本例使用 dlib 的 Struct SVM 学习一个简单的多类分类器的参数。首先实现多类分类器模型，然后利用结构支持向量机工具进行遍历，找到该分类模型的参数。实例文件 svm_struct.py 的具体实现代码如下。

源码路径：daima\4\svm_struct.py

(1) 在本例中创建 3 种类型的示例：类 0、1 或 2。也就是说，我们的每个样本向量分为 3 类。为了使这个例子非常简单，除了一个地方外，每个样本向量在任何地方都是零。每个向量的非零维数决定了向量的类别。例如，samples 的第一个元素的类为 1，因为 samples[0][1]是 samples[0]中唯一的非零元素。代码如下：

```
def main():
    samples = [[0, 2, 0], [1, 0, 0], [0, 4, 0], [0, 0, 3]]
    #由于我们想使用机器学习方法来学习 3 类分类器,所以我们需要记录样本的标签。这里的 samples[i]
    #有一个标签 labels[i]的类标签
    labels = [1, 0, 1, 2]
    problem = ThreeClassClassifierProblem(samples, labels)
    weights = dlib.solve_structural_svm_problem(problem)
```

(2) 打印权重信息，然后对每个训练样本计算 predict_label()，请注意：每个样本都预测了正确的标签。代码如下：

```
    print(weights)
    for k, s in enumerate(samples):
        print("Predicted label for sample[{0}]: {1}".format(
            k, predict_label(weights, s)))
```

(3) 创建函数 predict_label(weights, sample)，设置 3 类分类器的 9 维权向量，预测指定 3 维样本向量的类。因此，此函数的输出为 0、1 或 2(即 3 个可能的标签之一)。我们可以将 3 类分类器模型看作是包含 3 个独立的线性分类器。因此，为了预测样本向量的类别，需要评估这 3 个分类器中的每一个。下面的代码只是从权重中提取 3 个独立的权重向量，然后根据样本对每个向量求值。单个分类器的得分存储在得分中最高得分索引作为标签返回。

代码如下：

```
def predict_label(weights, sample):
   w0 = weights[0:3]
   w1 = weights[3:6]
   w2 = weights[6:9]
   scores = [dot(w0, sample), dot(w1, sample), dot(w2, sample)]
   max_scoring_label = scores.index(max(scores))
   return max_scoring_label
```

(4) 编写函数 dot(a, b)，计算两个向量 a 和 b 之间的点积。代码如下：

```
def dot(a, b):
   return sum(i * j for i, j in zip(a, b))
```

(5) 定义类 ThreeClassClassifierProblem，使用 dlib.solve_structural_svm_problem()告诉 Struct SVM 多分类器如何处理我们的问题。Struct SVM 支持向量机是一种有监督的机器学习方法，用于学习预测复杂的输出。这与只做简单的“是/否”预测的二元分类器形成了对比。另一方面，是/否向量机可以学习预测复杂的输出，例如整个解析树或 DNA 序列比对。为此，它学习一个函数 F(x, y)，该函数测量特定数据样本 x 与标签 y 的匹配程度，其中标签可能是一个复杂的东西，比如解析树。然而，为了使本实例程序尽量简单，我们只使用了 3 个类别的标签输出。在测试时，新 x 的最佳标签由最大化 F(x, y)的 y 给出，把它放到当前示例的上下文中，F(x, y)能够计算给定样本和类标签的分数。因此，预测的类标签是使 F(x, y)最大的 y 的任何值。这正是 predict_u label()所做的。也就是说，它计算 F(x, 0)、F(x, 1)和 F(x, 2)，然后报告哪个标签的值最大。代码如下：

```
class ThreeClassClassifierProblem:

   def __init__(self, samples, labels):
      # dlib.solve_structural_svm_problem() 要求类具有 num_samples 和 num_dimensions
      # 字段。这些字段应分别包含训练样本数和 PSI 特征向量的维数
      self.num_samples = len(samples)
      self.num_dimensions = len(samples[0])*3

      self.samples = samples
      self.labels = labels
```

(6) 编写函数 make_psi(self, x, label)计算 PSI(x,label)，在这里所做的就是取 x。在本实例程序中是一个三维样本向量，把它放到一个 9 维 PSI 向量的 3 个位置之一，然后返回，所以函数 make_psi()返回 PSI(x，label)。要了解为什么要这样设置 PSI，请回想一下 predict_u label()的工作原理，它接受一个 9 维的权重向量，并将向量分成 3 部分。然后每个片段定义

一个不同的分类器，我们用一对一的方式使用它们来预测标签。所以，现在在 Struct SVM 向量机代码中，我们必须定义 PSI 向量来对应这个用法。也就是说，我们需要告诉结构支持向量机解算器我们要解决什么样的问题。代码如下：

```
def make_psi(self, x, label):

    psi = dlib.vector()
    # 设置有 9 个维度,向量元素初始化为 0
    psi.resize(self.num_dimensions)
    dims = len(x)
    if label == 0:
        for i in range(0, dims):
            psi[i] = x[i]
    elif label == 1:
        for i in range(dims, 2 * dims):
            psi[i] = x[i - dims]
    else:  # 标签必须为 2
        for i in range(2 * dims, 3 * dims):
            psi[i] = x[i - 2 * dims]
    return psi
```

(7) 编写 dlib 直接调用的两个成员函数 get_truth_joint_feature_vector(self, idx)和 separation_oracle(self, idx, current_solution)。在 get_truth_joint_feature_vector()中，当第 idx 个训练样本具有真实标签时，只需返回该样本的 PSI()向量，所以这里它会返回 PSI(self.samples[idx], self.labels[idx])。代码如下：

```
def get_truth_joint_feature_vector(self, idx):
    return self.make_psi(self.samples[idx], self.labels[idx])

def separation_oracle(self, idx, current_solution):
    samp = self.samples[idx]
    dims = len(samp)
    scores = [0, 0, 0]
    # 计算 3 个分类器的得分
    scores[0] = dot(current_solution[0:dims], samp)
    scores[1] = dot(current_solution[dims:2*dims], samp)
    scores[2] = dot(current_solution[2*dims:3*dims], samp)

    if self.labels[idx] != 0:
        scores[0] += 1
    if self.labels[idx] != 1:
        scores[1] += 1
    if self.labels[idx] != 2:
        scores[2] += 1
```

```
        max_scoring_label = scores.index(max(scores))
        if max_scoring_label == self.labels[idx]:
            loss = 0
        else:
            loss = 1

        # 最后，返回刚刚找到的标签对应的损失和 PSI 向量
        psi = self.make_psi(samp, max_scoring_label)
        return loss, psi


if __name__ == "__main__":
    main()
```

执行后会输出：

```
0.25
-0.166665
-0.111114
-0.125011
0.333332
-0.111111
-0.124989
-0.166667
0.222225
Predicted label for sample[0]: 1
Predicted label for sample[1]: 0
Predicted label for sample[2]: 1
Predicted label for sample[3]: 2
```

4.4 自训练模型

在本章前面的内容中，我们使用的都是从 dlib 官方下载的数据集模型。通过使用机器学习技术，我们可以开发出自己的模型数据集，可以使用 dlib 机器学习工具训练自己的模型，也可以制作自己的对象检测器。

扫码观看本节视频讲解

4.4.1 训练自己的模型

通过使用 dlib 提供的机器学习工具，开发者可以训练自己的模型。请看下面的实例文

件 train_shape_predictor.py，功能是训练一个基于小数据集的人脸标记模型，然后对其进行评估。本实例使用了 dlib 论文中的算法《One Millisecond Face Alignment with an Ensemble of Regression Trees，Vahid Kazemi 和 Josephine Sullivan，CVPR 2014》。如果要在某些图像上进行可视化训练模型的输出，则可以运行实例文件 train_shape_predictor.py，并将文件 predictor.dat 作为输入模型。需要注意的是，此类模型虽然经常用于人脸标记领域，但是却非常通用，也可以用于各种形状的预测任务。但是在本实例中，我们仅使用一个简单的面部标记任务应用来演示训练自己模型的知识。实例文件 train_shape_predictor.py 的具体实现流程如下。

源码路径：daima\4\train_shape_predictor.py

(1) 在本实例中将基于 examples/faces 目录中的 small faces 数据集训练一个人脸检测器，需要提供此 faces 文件夹的路径作为命令行参数，以便让程序知道它在哪里。

```
if len(sys.argv) != 2:
    print(
        "Give the path to the examples/faces directory as the argument to this "
        "program. For example, if you are in the python_examples folder then "
        "execute this program by running:\n"
        "    ./train_shape_predictor.py ../examples/faces")
    exit()
faces_folder = sys.argv[1]

options = dlib.shape_predictor_training_options()
```

(2) 现在开始训练模型。在本算法中可以设置多个参数，在 shape_predictor_trainer 的文档中说明了这些参数的含义，在 Kazemi 的论文中也详细解释了各个参数。在本实例中，只是设置了其中的 3 个参数的值，没有使用这 3 个参数的默认值。这样设置的原因是，本实例用到的只是一个非常小的数据集。其中将采样参数设置为较高的值(300)，可以有效地提高训练集的大小。

```
options.oversampling_amount = 300
#增加正则化(使 nu 变小)和使用深度更小的树来降低模型的容量
options.nu = 0.05
options.tree_depth = 2
options.be_verbose = True
```

(3) 使用函数 dlib.train_shape_predictor()做实际训练工作，将最终的预测器数据保存到文件 predictor.dat 中。输入的是一个 XML 文件，在其中列出了训练数据集中的图像，也包含了面部各个成员的位置。

```
training_xml_path = os.path.join(faces_folder, "training_with_face_landmarks.xml")
dlib.train_shape_predictor(training_xml_path, "predictor.dat", options)
```

(4) 现在已经有了一个数据模型，接下来就可以测试这个模型。函数 test_shape_predictor()用于测量 shape_predictor 输出的人脸标记与根据真值数据应该位于的位置之间的平均距离。

```
print("\n 训练精度: {}".format(
   dlib.test_shape_predictor(training_xml_path, "predictor.dat")))
```

(5) 在真正的测试工作中，看看它在没有经过训练的数据上做得有多好。因为我们在一个非常小的数据集上进行训练，所以精确度不是很高，但是效果还是很满意的。此外，如果在一个大的人脸标记数据集上训练它，可以获得最好的结果，正如 Kazemi 的论文所示。

```
testing_xml_path = os.path.join(faces_folder,
"testing_with_face_landmarks.xml")
print("测试精度 {}".format(
   dlib.test_shape_predictor(testing_xml_path, "predictor.dat")))
```

(6) 像在普通应用程序中一样使用预测数据 predictor.dat，首先从磁盘中加载预测数据文件 predictor.dat，然后还需要加载一个人脸检测器来提供人脸位置的初始估计值。

```
predictor = dlib.shape_predictor("predictor.dat")
detector = dlib.get_frontal_face_detector()
```

(7) 对 faces 文件夹中的图像运行 detector 和 shape_predictor 并显示结果。

```
print("显示 faces 文件夹中图像的检测和预测...")
win = dlib.image_window()
for f in glob.glob(os.path.join(faces_folder, "*.jpg")):
   print("正在处理文件: {}".format(f))
   img = dlib.load_rgb_image(f)

   win.clear_overlay()
   win.set_image(img)
```

(8) 让探测器找到每个面部的边界框，其中第 3 个参数 1 表示将图像向上采样一次，这样可以让我们能够发现更多的面孔。

```
   dets = detector(img, 1)
   print("检测到的面数: {}".format(len(dets)))
   for k, d in enumerate(dets):
      print("检测 {}: Left: {} Top: {} Right: {} Bottom: {}".format(
         k, d.left(), d.top(), d.right(), d.bottom()))
      #在方框 d 中获取面部的 landmarks/parts
      shape = predictor(img, d)
```

```
        print("Part 0: {}, Part 1: {} ...".format(shape.part(0),
                                                   shape.part(1)))
        #在屏幕中绘制出面部标志。
        win.add_overlay(shape)

    win.add_overlay(dets)
    dlib.hit_enter_to_continue()
```

输入下面的命令运行本实例，运行代码后会创建自制的数据集文件 predictor.dat，并检测 examples/faces 目录中各个图片的人脸，如图 4-11 所示。每当按回车键，可以逐一检测 examples/faces 目录中的图片。

```
python train_shape_predictor.py faces
```

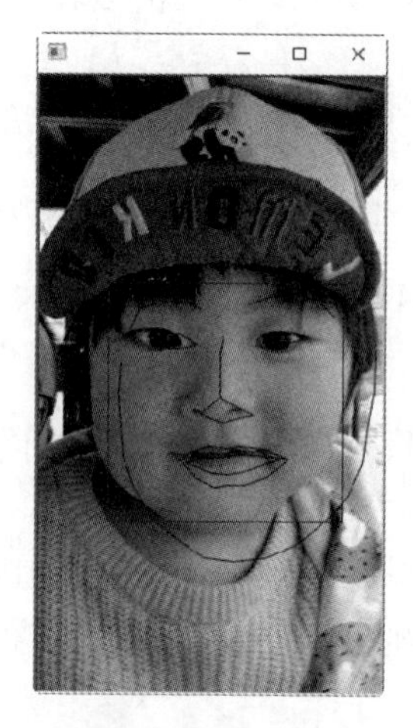

图 4-11　执行效果

4.4.2 自制对象检测器

通过使用 dlib 提供的机器学习工具，开发者可以制作自己的人脸对象检测器。请看下面的实例文件 train_object_detector.py，功能是使用 dlib 为人脸、行人和任何其他半刚性对象制作基于 HOG 的对象检测器。本实例参考了 Dalal 和 Triggs 在 2005 年发表的《面向人类检测的梯度直方图》论文，在论文中首次提出了滑动窗口目标检测器的训练步骤。实例文件 train_object_detector.py 的具体实现流程如下。

源码路径：daima\4\train_object_detector.py

(1) 基于 examples/faces 目录中的 small faces 数据集训练一个人脸检测器，代码如下：

```
if len(sys.argv) != 2:
    print(
        "Give the path to the examples/faces directory as the argument to this "
        "program. For example, if you are in the python_examples folder then "
        "execute this program by running:\n"
        "    ./train_object_detector.py ../examples/faces")
    exit()
faces_folder = sys.argv[1]
```

(2) 开始训练。函数 simple_object_detector_training_options()有一系列 options 选项，所有选项都有自带的默认值。常用 options 的具体说明如下。

- add_left_right_image_flips：因为人脸是左右对称的，所以我们可以设置此选项为 True。告诉训练一个对称的检测器，这有助于从训练数据中获得最大的价值。
- c：因为训练器是一种向量机，因此通常具有的 svmc 参数。一般来说，c 越大就越适合训练数据，但可能会导致过度拟合。必须检查经过训练的检测器在未训练的图像测试集上的工作情况，根据经验找到 c 的最佳值。通常不建议把值设置为 5。请尝试设置几个不同的 c 值，看看哪个最适合我们的数据。
- num_threads：告诉代码当前计算机有多少个 CPU 核心来进行最快的训练。

代码如下：

```
options = dlib.simple_object_detector_training_options()
options.add_left_right_image_flips = True
options.C = 5
options.num_threads = 4
options.be_verbose = True


training_xml_path = os.path.join(faces_folder, "training.xml")
testing_xml_path = os.path.join(faces_folder, "testing.xml")
```

(3) 函数 train_simple_object_detector()用于实际训练工作，将最终探测器保存到文件 detector.svm 中。输入参数是一个 XML 文件，它列出了训练数据集中的图像，还包含了面框的位置。如果开发者想要创建自己的 XML 文件，可以在 tools/imglab 文件夹中找到 imglab 工具。这是一个简单的图形工具，用于使用方框标记图像中的对象。要了解如何使用它，请阅读帮助文档 tools/imglab/README.txt。对于本实例来说，我们只使用 dlib 中包含的文件 training.xml。代码如下：

```
dlib.train_simple_object_detector(training_xml_path, "detector.svm", options)
```

(4) 现在已经有了人脸探测器，接下来就可以进行测试了。在训练数据上测试我们的人脸探测器，打印输出平均精度信息。代码如下：

```
print("")                          #打印空行以创建与上一输出之间的间隙
print("Training accuracy: {}".format(
   dlib.test_simple_object_detector(training_xml_path, "detector.svm")))
```

(5) 要想知道探测器是否真的在不过度拟合的情况下工作，需要在没有经过训练的图像上运行。事实会证明，我们的探测器将在测试图像上完美地工作。代码如下：

```
print("Testing accuracy: {}".format(
   dlib.test_simple_object_detector(testing_xml_path, "detector.svm")))
```

(6) 像在正常应用中一样使用探测器，首先从磁盘加载探测器，代码如下：

```
detector = dlib.simple_object_detector("detector.svm")

# 可以看看我们学过的 HOG 过滤器，看起来像一张脸
win_det = dlib.image_window()
win_det.set_image(detector)
```

(7) 使用探测器检测文件夹 faces 中的图像，并打印输出检测结果。代码如下：

```
print("Showing detections on the images in the faces folder...")
win = dlib.image_window()
for f in glob.glob(os.path.join(faces_folder, "*.jpg")):
   print("Processing file: {}".format(f))
   img = dlib.load_rgb_image(f)
   dets = detector(img)
   print("Number of faces detected: {}".format(len(dets)))
   for k, d in enumerate(dets):
      print("Detection {}: Left: {} Top: {} Right: {} Bottom: {}".format(
         k, d.left(), d.top(), d.right(), d.bottom()))

   win.clear_overlay()
   win.set_image(img)
   win.add_overlay(dets)
   dlib.hit_enter_to_continue()
```

(8) 假设已经训练了多个探测器。并且希望将它们作为一个组高效地运行，我们可以按照以下的方式执行此操作。代码如下：

```
detector1 = dlib.fhog_object_detector("detector.svm")
```

(9) 再次加载探测器文件 detector.svm，代码如下：

```
detector2 = dlib.fhog_object_detector("detector.svm")
```

(10) 列出所有想运行的探测器。现在我们有两个探测器，可以用任何数字命名。代码如下：

```
detectors = [detector1, detector2]
image = dlib.load_rgb_image(faces_folder + '/2008_002506.jpg')
[boxes, confidences, detector_idxs] = dlib.fhog_object_detector.run_multiple
            (detectors, image, upsample_num_times=1, adjust_threshold=0.0)
for i in range(len(boxes)):
   print("detector {} found box {} with confidence {}.".format(detector_idxs[i],
         boxes[i], confidences[i]))
```

(11) 请注意，不必使用基于 XML 的输入来训练函数 simple_object_detector()，如果已经加载了探测目标对象的训练图像和边界框，则可以通过如下代码进行调用，在调用时只需把图片放到一个列表中即可。代码如下：

```
images = [dlib.load_rgb_image(faces_folder + '/2008_002506.jpg'),
       dlib.load_rgb_image(faces_folder + '/2009_004587.jpg')]
# 为每个图像制作一个矩形列表，给出框边的像素位置
boxes_img1 = ([dlib.rectangle(left=329, top=78, right=437, bottom=186),
           dlib.rectangle(left=224, top=95, right=314, bottom=185),
           dlib.rectangle(left=125, top=65, right=214, bottom=155)])
boxes_img2 = ([dlib.rectangle(left=154, top=46, right=228, bottom=121),
           dlib.rectangle(left=266, top=280, right=328, bottom=342)])
# 将这些框列表聚合为一个大列表，然后调用 train_simple_object_detector()
boxes = [boxes_img1, boxes_img2]

detector2 = dlib.train_simple_object_detector(images, boxes, options)
# 可以通过取消注释以下内容将此探测器保存到磁盘
#detector2.save('detector2.svm')

# 现在看看它的 HOG 过滤器
win_det.set_image(detector2)
dlib.hit_enter_to_continue()

# 请注意，不必使用基于 XML 的输入来测试 simple_object_detector()
# 如果已经加载了对象的训练图像和边界框，则可以通过如下代码调用
print("\nTraining accuracy: {}".format(
   dlib.test_simple_object_detector(images, boxes, detector2)))
```

输入下面的命令运行本实例，运行代码后会创建自制的探测器文件 detector.svm，测试效果如图 4-12 所示。并检测 examples/faces 目录中各个图片的人脸，如图 4-13 所示。每当按回车键，可以逐一检测 examples/faces 目录中的图片。

```
python train_object_detector.py faces
```

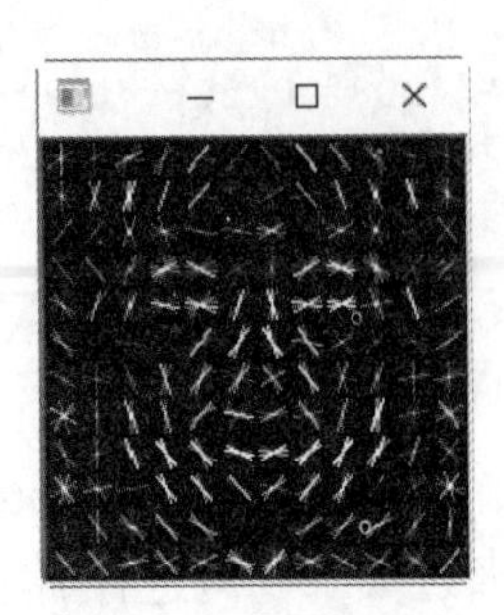

图 4-12　探测器效果

图 4-13　测试效果

第5章 face_recognition 人脸识别

face_recognition 是基于 C++开源库 dlib 实现的，是一个强大、简单、易上手的人脸识别开源项目，并且配备了完整的开发文档和应用案例。通过使用 face_recognition，可以在 Python 程序中实现人脸识别功能。在本章的内容中，将详细讲解使用 face_recognition 实现人脸识别的知识。

5.1 安装 face_recognition

在安装face_recongnition之前，我们必须明白如下的几个依赖关系。

- 安装 face_recongnition 的必要条件是：配置好库 dlib。
- 配置好 dlib 的必要条件是：成功安装 dlib，并且编译。
- 安装 dlib 的必要条件是：配置好 boost 和 cmake。

扫码观看本节视频讲解

在 Python 3.6 之前的版本中，开发者必须严格按照上述依赖关系进行环境搭建。而从 Python 3.6 开始，安装 face_recognition 非常容易，整个过程与 boost 和 cmake 完全无关。但是我们必须注意的是，dlib 针对不同的 Python 版本提供了不同的安装文件，我们必须安装完全对应的版本，否则会出错。例如笔者安装的是 Python 3.6，所以必须安装的 dlib 版本是 19.7.0。下载 dlib-19.7.0-cp36-cp36m-win_amd64.whl 后，使用如下所示的命令即可成功安装 dlib：

```
pip install dlib-19.7.0-cp36-cp36m-win_amd64.whl
```

接下来通过如下命令即可安装 face_recognition：

```
pip install face_recognition
```

除此之外，还需要安装如下所示的库：

```
pip install numpy
pip install scipy
pip install opencv-python
```

到此为止，在 Python 环境中安装库 face_recognition 的工作结束。

5.2 实现基本的人脸检测

在本节的内容中，将通过具体实例的实现过程，详细讲解使用 face_recognition 实现基本人脸检测的知识。

扫码观看本节视频讲解

5.2.1 输出显示指定人像人脸特征

库 face_recognition 通过 facial_features 来处理面部特征，包含了如

下 8 个特征：

- chin：下巴。
- left_eyebrow：左眉。
- right_eyebrow：右眉。
- nose_bridge：鼻梁。
- nose_tip：鼻子尖。
- left_eye：左眼。
- right_eye：右眼。
- top_lip：上唇。
- bottom_lip：下唇。

例如在下面的实例文件 shibie01.py 中，演示了输出显示指定人像人脸特征的过程。

源码路径：daima\5\5-1\shibie01.py

```
# 自动识别人脸特征

# 导入pil 模块
from PIL import Image, ImageDraw
# 导入face_recogntion 模块，可用命令安装 pip install face_recognition
import face_recognition

# 将jpg 文件加载到 Numpy 数组中
image = face_recognition.load_image_file("111.jpg")

#查找图像中所有的面部特征
face_landmarks_list = face_recognition.face_landmarks(image)

print("I found {} face(s) in this
photograph.".format(len(face_landmarks_list)))

for face_landmarks in face_landmarks_list:

   #打印此图像中每个面部特征的位置
    facial_features = [
        'chin',
        'left_eyebrow',
        'right_eyebrow',
        'nose_bridge',
        'nose_tip',
        'left_eye',
        'right_eye',
        'top_lip',
```

```
    'bottom_lip'
]

for facial_feature in facial_features:
    print("The {} in this face has the following points: {}".format
          (facial_feature, face_landmarks[facial_feature]))

#在图像中描绘出每个人脸特征
pil_image = Image.fromarray(image)
d = ImageDraw.Draw(pil_image)

for facial_feature in facial_features:
    d.line(face_landmarks[facial_feature], width=5)

pil_image.show()
```

执行代码后会输出显示图片 111.jpg 中人像的人脸特征数值：

```
I found 1 face(s) in this photograph.
The chin in this face has the following points: [(35, 303), (38, 331), (42, 359), (47, 387), (59, 411), (78, 428), (100, 441), (123, 451), (146, 454), (168, 450), (189, 439), (209, 425), (227, 407), (238, 384), (244, 358), (249, 331), (252, 304)]
The left_eyebrow in this face has the following points: [(53, 289), (66, 273), (87, 266), (109, 269), (131, 276)]
The right_eyebrow in this face has the following points: [(162, 277), (181, 269), (203, 266), (224, 271), (236, 286)]
The nose_bridge in this face has the following points: [(144, 303), (144, 319), (144, 334), (143, 351)]
The nose_tip in this face has the following points: [(124, 364), (134, 366), (144, 368), (154, 366), (164, 364)]
The left_eye in this face has the following points: [(77, 304), (89, 299), (103, 300), (115, 310), (102, 313), (87, 311)]
The right_eye in this face has the following points: [(174, 310), (185, 301), (199, 301), (211, 305), (200, 312), (186, 313)]
The top_lip in this face has the following points: [(109, 395), (124, 391), (136, 387), (144, 389), (153, 386), (165, 390), (180, 393), (174, 393), (153, 394), (144, 395), (136, 394), (116, 396)]
The bottom_lip in this face has the following points: [(180, 393), (165, 402), (154, 405), (145, 406), (136, 406), (125, 404), (109, 395), (116, 396), (136, 396), (145, 396), (153, 395), (174, 393)]
```

并且会使用 PIL 在人像中标记出人脸特征，如图 5-1 所示。

图 5-1　标记出人脸特征

5.2.2　在指定照片中识别标记出人脸

在下面的实例文件 shibie.py 中，演示了在指定照片中识别标记出人脸的过程。

源码路径：daima\5\5-2\shibie.py

```
# 检测人脸
import face_recognition
import cv2

# 读取图片并识别人脸
img = face_recognition.load_image_file("111.jpg")
face_locations = face_recognition.face_locations(img)
print(face_locations)

# 调用 opencv 函数显示图片
img = cv2.imread("111.jpg")
cv2.namedWindow("原图")
cv2.imshow("原图", img)

# 遍历每个人脸，并标注
faceNum = len(face_locations)
for i in range(0, faceNum):
    top =  face_locations[i][0]
    right =  face_locations[i][1]
    bottom = face_locations[i][2]
```

```
    left = face_locations[i][3]

    start = (left, top)
    end = (right, bottom)

    color = (55,255,155)
    thickness = 3
    cv2.rectangle(img, start, end, color, thickness)

# 显示识别结果
cv2.namedWindow("识别")
cv2.imshow("识别", img)
cv2.waitKey(0)
cv2.destroyAllWindows()
```

执行代码后将分别显示原始照片 111.jpg 效果和识别标记处人脸的效果，如图 5-2 所示。

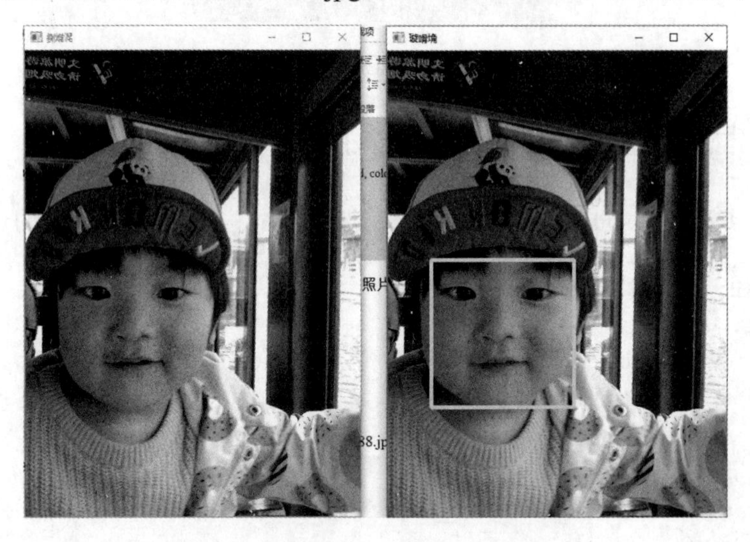

图 5-2　原始照片效果和标记出人脸效果

5.2.3　识别出照片中的所有人脸

假设我们有一张照片 888.jpg，如图 5-3 所示。

这是一幅 3 人合影照，我们应该如何识别出这张照片中的人脸呢？在下面的实例文件 shibie02.py 中，演示了提取出照片 888.jpg 中所有人脸的过程。

图 5-3 照片 888.jpg

源码路径：daima\5\5-3\shibie02.py

```
#  识别图片中的所有人脸并显示出来
# filename : shibie02.py
# 导入 pil 模块
from PIL import Image
# 导入 face_recogntion 模块，可用命令安装 pip install face_recognition
import face_recognition

# 将 jpg 文件加载到 Numpy 数组中
image = face_recognition.load_image_file("888.jpg")

# 使用默认的 HOG 模型查找图像中所有人脸
# 这个方法已经相当准确了，但还是不如 CNN 模型那么准确，因为没有使用 GPU 加速
# 另请参见：find_faces_in_picture_cnn.py
face_locations = face_recognition.face_locations(image)

# 使用 CNN 模型
# face_locations = face_recognition.face_locations(image,
number_of_times_to_upsample=0, model="cnn")

# 打印：我从图片中找到了多少张人脸
print("I found {} face(s) in this photograph.".format(len(face_locations)))

# 循环找到所有人脸
for face_location in face_locations:
```

```
        # 打印每张脸的位置信息
        top, right, bottom, left = face_location
        print("Top: {}, Left: {}, Bottom: {}, Right: {}".format(top, left, bottom,
              right))
# 指定人脸的位置信息，然后显示人脸图片
        face_image = image[top:bottom, left:right]
        pil_image = Image.fromarray(face_image)
        pil_image.show()
```

执行后首先会输出照片 888.jpg 中人脸的位置信息：

```
I found 3 face(s) in this photograph.
Top: 163, Left: 79, Bottom: 271, Right: 187
Top: 125, Left: 182, Bottom: 254, Right: 311
Top: 329, Left: 104, Bottom: 403, Right: 179
```

然后输出显示识别的 3 张人脸，如图 5-4 所示。

图 5-4　提取出的 3 张人脸

接下来再在下面的实例文件 shibie02-1.py 中，演示提取出照片 888.jpg 中所有人脸的一个过程。

源码路径：daima\5\5-4\shibie02-1.py

```
import face_recognition
import cv2

# 读取图片并识别人脸
img = face_recognition.load_image_file("888.jpg")
face_locations = face_recognition.face_locations(img)
print(face_locations)

# 调用 opencv 函数显示图片
img = cv2.imread("888.jpg")
cv2.namedWindow("原图")
cv2.imshow("原图", img)

# 遍历每个人脸并标注
faceNum = len(face_locations)
for i in range(0, faceNum):
```

```
        top = face_locations[i][0]
        right = face_locations[i][1]
        bottom = face_locations[i][2]
        left = face_locations[i][3]

        start = (left, top)
        end = (right, bottom)

        color = (55, 255, 155)
        thickness = 3
        cv2.rectangle(img, start, end, color, thickness)

# 显示识别结果
cv2.namedWindow("识别")
cv2.imshow("识别", img)

cv2.waitKey(0)
cv2.destroyAllWindows()
```

执行代码后会输出显示照片 888.jpg 中的所有人脸信息，如图 5-5 所示。

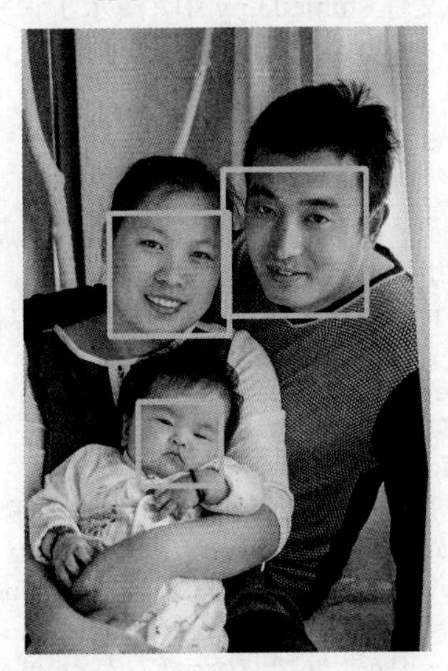

图 5-5　识别出的 3 张人脸

5.2.4　判断在照片中是否包含某个人脸

假设我们有一张照片 201.jpg，如图 5-6 所示。

图 5-6　照片 201.jpg

这是一幅单人照，假设这个人的名字叫小毛毛。请问我们应该如何识别出在照片 888.jpg 中有小毛毛呢？在下面的实例文件 shibie03.py 中，演示了识别判断在照片 888.jpg 中是否包含小毛毛人脸的过程。

源码路径：daima\5\5-5\shibie03.py

```
# 识别人脸，鉴定是哪个人
import face_recognition
#将 jpg 文件加载到 Numpy 数组中
chen_image = face_recognition.load_image_file("201.jpg")
#要识别的图片
unknown_image = face_recognition.load_image_file("888.jpg")
#获取每个图像文件中每个面部的面部编码
#由于每个图像中可能有多个面，所以返回一个编码列表
#但是由于知道每个图像只有一张脸，只关心每个图像中的第一个编码，所以取索引 0
chen_face_encoding = face_recognition.face_encodings(chen_image)[0]
print("chen_face_encoding:{}".format(chen_face_encoding))
unknown_face_encoding = face_recognition.face_encodings(unknown_image)[0]
print("unknown_face_encoding :{}".format(unknown_face_encoding))

known_faces = [
    chen_face_encoding
]
#结果是 True/False 的数组，未知面孔 known_faces 阵列中是任何人相匹配的结果
results = face_recognition.compare_faces(known_faces, unknown_face_encoding)
```

```
print("result :{}".format(results))
print("这个未知面孔是 小毛毛 吗？{}".format(results[0]))
print("这个未知面孔是 我们从未见过的新面孔吗？{}".format(not True in results))
```

执行代码后会输出如下识别结果，这说明在照片 888.jpg 中存在照片 201.jpg 的这个人。

```
result :[True]
这个未知面孔是 小毛毛 吗？True
这个未知面孔是 我们从未见过的新面孔吗？False
```

5.2.5 识别出在照片中的人到底是谁

假设分别存在 3 张图片 laoguan.jpg(老管的单人照)、maomao.jpg(毛毛的单人照)和 unknown.jpg(未知某人的单人照，肯定是老管或毛毛这两人之一)，如图 5-7 所示。

laoguan.jpg

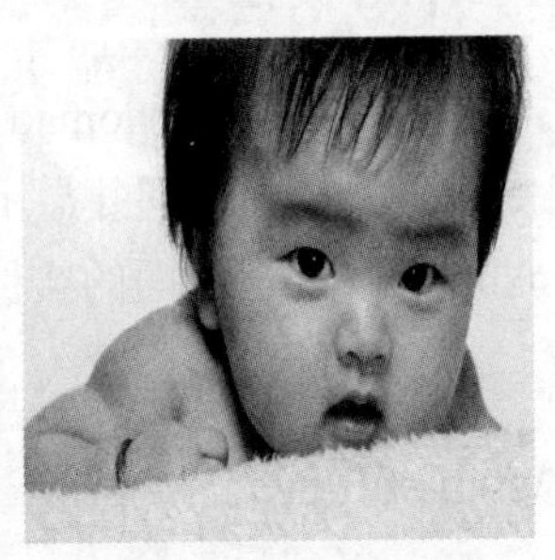

maomao.jpg

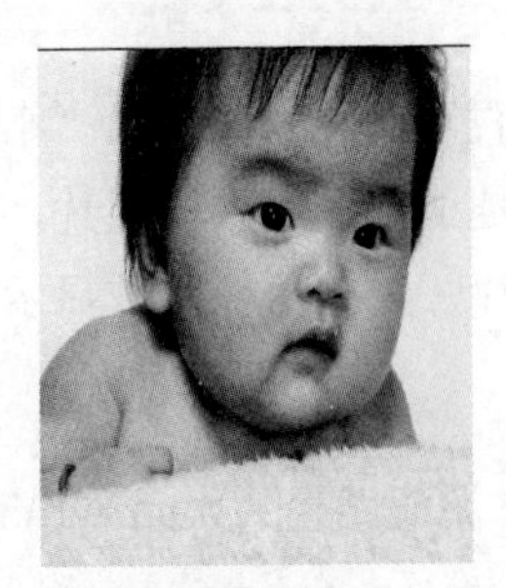

unknown.jpg

图 5-7 3 张素材图片

我们应该如何识别出照片 unknown.jpg 中的人是谁呢？在下面的实例文件 shibie04.py 中，演示了识别判断在照片 unknown.jpg 中的人到底是谁的过程。

源码路径：daima\5\5-6\shibie04.py

```
# 识别图片中的人脸
import face_recognition
jobs_image = face_recognition.load_image_file("laoguan.jpg");
obama_image = face_recognition.load_image_file("maomao.jpg");
unknown_image = face_recognition.load_image_file("unknown.jpg");

laoguan_encoding = face_recognition.face_encodings(jobs_image)[0]
maomao_encoding = face_recognition.face_encodings(obama_image)[0]
unknown_encoding = face_recognition.face_encodings(unknown_image)[0]
```

```
results = face_recognition.compare_faces([laoguan_encoding, maomao_encoding],
                                          unknown_encoding )
labels = ['老管', '毛毛']

print('结果:'+str(results))

for i in range(0, len(results)):
    if results[i] == True:
        print('这个人是:'+labels[i])
```

执行代码后会成功输出识别结果：

```
结果:[False, True]
这个人是:毛毛
```

5.2.6 摄像头实时识别

假设我们保存一张“小毛毛”的照片 xiaomaomao.jpg，然后用摄像头识别不同的照片，如果是小毛毛本人的照片，则摄像区域自动识别并显示“小毛毛”。如果摄像头不是小毛毛的照片，则在摄像区域显示 unknown。通过下面的实例文件 shibie05.py 可以实现上述识别功能。

源码路径：daima\5\5-7\shibie05.py

```
import face_recognition
import cv2

video_capture = cv2.VideoCapture(0)#笔记本摄像头是 0，外接摄像头设备是 1

obama_img = face_recognition.load_image_file("xiaomaomao.jpg")
obama_face_encoding = face_recognition.face_encodings(obama_img)[0]

face_locations = []
face_encodings = []
face_names = []
process_this_frame = True

while True:
    ret, frame = video_capture.read()

    small_frame = cv2.resize(frame, (0, 0), fx=0.25, fy=0.25)
    if process_this_frame:
        face_locations = face_recognition.face_locations(small_frame)
```

```
        face_encodings = face_recognition.face_encodings(small_frame,
                                                          face_locations)

        face_names = []
        for face_encoding in face_encodings:
            match = face_recognition.compare_faces([obama_face_encoding],
                                                   face_encoding)

            if match[0]:
                name = "小毛毛"
            else:
                name = "unknown"

            face_names.append(name)

    process_this_frame = not process_this_frame

    for (top, right, bottom, left), name in zip(face_locations, face_names):
        top *= 4
        right *= 4
        bottom *= 4
        left *= 4

        cv2.rectangle(frame, (left, top), (right, bottom), (0, 0, 255),  2)

        cv2.rectangle(frame, (left, bottom - 35), (right, bottom), (0, 0, 255), 2)
        font = cv2.FONT_HERSHEY_DUPLEX
        cv2.putText(frame, name, (left+6, bottom-6), font, 1.0, (255, 255, 255), 1)

    cv2.imshow('Video', frame)

    if cv2.waitKey(1) & 0xFF == ord('q'):
        break

video_capture.release()
cv2.destroyAllWindows()
```

再看下面的实例文件 facerec_from_webcam.py，功能是识别摄像头中的人脸。本实例使用库 OpenCV 从摄像头读取视频。首先准备两幅素材图片 obama.jpg 和 biden.jpg，然后识别摄像头中的人脸。如果摄像头中的人脸是两幅素材图片 obama.jpg 和 biden.jpg 的人脸，则在摄像头视频中用矩形标签注明识别结果。如果摄像头中的人脸不是两幅素材图片 obama.jpg 和 biden.jpg 的人脸，则在摄像头视频中用矩形标签注明 Unknown。

源码路径：daima\5\5-8\facerec_from_webcam.py

```
import face_recognition
import cv2
import numpy as np

video_capture = cv2.VideoCapture(0)

#加载实例图片并学习如何识别它
obama_image = face_recognition.load_image_file("obama.jpg")
obama_face_encoding = face_recognition.face_encodings(obama_image)[0]

#加载第二个实例图片并学习如何识别它。
biden_image = face_recognition.load_image_file("biden.jpg")
biden_face_encoding = face_recognition.face_encodings(biden_image)[0]

#创建已知面编码及其名称的数组
known_face_encodings = [
    obama_face_encoding,
    biden_face_encoding
]
known_face_names = [
    "Barack Obama",
    "Joe Biden"
]

while True:
    #抓取一帧视频
    ret, frame = video_capture.read()

    #将图像从BGR颜色(OpenCV使用)转换为RGB颜色(人脸识别可以使用的颜色)
    rgb_frame = frame[:, :, ::-1]

    #找到视频帧中的所有人脸和人脸编码
    face_locations = face_recognition.face_locations(rgb_frame)
    face_encodings = face_recognition.face_encodings(rgb_frame,
face_locations)

    #在这一帧视频中遍历每个人脸
    for (top, right, bottom, left), face_encoding in zip(face_locations,
                                                          face_encodings):
        #查看该脸是否与已知人脸匹配
        matches = face_recognition.compare_faces(known_face_encodings,
                                                 face_encoding)

        name = "Unknown"
```

```
            #或者使用与新人脸距离最小的已知人脸
            face_distances = face_recognition.face_distance(known_face_encodings,
                                                            face_encoding)
            best_match_index = np.argmin(face_distances)
            if matches[best_match_index]:
                name = known_face_names[best_match_index]

            # 在人脸上画一个方框
            cv2.rectangle(frame, (left, top), (right, bottom), (0, 0, 255), 2)

            # 在人脸下方绘制一个带有名称的标签
            cv2.rectangle(frame, (left, bottom - 35), (right, bottom), (0, 0, 255),
                          cv2.FILLED)
            font = cv2.FONT_HERSHEY_DUPLEX
            cv2.putText(frame, name, (left + 6, bottom - 6), font, 1.0, (255, 255,
                        255), 1)

        #显示结果图像
        cv2.imshow('Video', frame)

        #按键盘中的 q 键退出
        if cv2.waitKey(1) & 0xFF == ord('q'):
            break

    #释放资源
    video_capture.release()
    cv2.destroyAllWindows()
```

执行代码后的效果如图 5-8 所示。

图 5-8 执行效果

上述实例文件 facerec_from_webcam_faster.py 的效率低，比较消耗计算机内存资源。在下面的实例文件中，我们对基本的识别功能进行了调整，使得识别速度更加快速。

(1) 以 1/4 分辨率处理每个视频帧，仍以全分辨率显示。

(2) 每隔一帧视频检测人脸。

源码路径：daima\5\5-9\facerec_from_webcam_faster.py

```
import face_recognition
import cv2
import numpy as np

video_capture = cv2.VideoCapture(0)
#加载实例图片并学习如何识别它
obama_image = face_recognition.load_image_file("obama.jpg")
obama_face_encoding = face_recognition.face_encodings(obama_image)[0]

#加载第二个实例图片并学习如何识别它
biden_image = face_recognition.load_image_file("biden.jpg")
biden_face_encoding = face_recognition.face_encodings(biden_image)[0]

#创建已知面编码及其名称的数组
known_face_encodings = [
   obama_face_encoding,
   biden_face_encoding
]
known_face_names = [
   "Barack Obama",
   "Joe Biden"
]

#初始化一些变量
face_locations = []
face_encodings = []
face_names = []
process_this_frame = True

while True:
    #抓取一帧视频
    ret, frame = video_capture.read()
    #将视频帧调整为1/4大小以快速实现人脸识别
    small_frame = cv2.resize(frame, (0, 0), fx=0.25, fy=0.25)
    #将图像从BGR颜色(OpenCV使用)转换为RGB颜色(人脸识别使用)
    rgb_small_frame = small_frame[:, :, ::-1]
    #每隔一帧的处理视频以节省时间
    if process_this_frame:
    #查找当前视频帧中的所有面和面编码
```

```
            face_locations = face_recognition.face_locations(rgb_small_frame)
            face_encodings = face_recognition.face_encodings(rgb_small_frame,
                                                             face_locations)

            face_names = []
            for face_encoding in face_encodings:
                # 查看该脸是否与已知人脸匹配
                matches = face_recognition.compare_faces(known_face_encodings,
                                                         face_encoding)

                name = "Unknown"
                face_distances = face_recognition.face_distance(known_face_encodings,
                                                                face_encoding)
                best_match_index = np.argmin(face_distances)
                if matches[best_match_index]:
                    name = known_face_names[best_match_index]
                face_names.append(name)

        process_this_frame = not process_this_frame

        #显示结果
        for (top, right, bottom, left), name in zip(face_locations, face_names):
            #缩放脸部位置，因为检测到的帧已缩放为1/4大小
            top *= 4
            right *= 4
            bottom *= 4
            left *= 4

            # 在脸上画一个方框
            cv2.rectangle(frame, (left, top), (right, bottom), (0, 0, 255), 2)

            #在脸部下方绘制一个带有名称的标签
            cv2.rectangle(frame, (left, bottom - 35), (right, bottom), (0, 0, 255),
                          cv2.FILLED)
            font = cv2.FONT_HERSHEY_DUPLEX
            cv2.putText(frame, name, (left + 6, bottom - 6), font, 1.0, (255, 255,
                        255), 1)

        #显示结果图像
        cv2.imshow('Video', frame)

        #按键盘中的q键将退出
        if cv2.waitKey(1) & 0xFF == ord('q'):
            break

    video_capture.release()
    cv2.destroyAllWindows()
```

5.3 深入 face_recognition 人脸检测

在本章前面的内容中，已经讲解了使用 face_recognition 实现基本人脸检测的知识。在本节的内容中，将进一步深入讲解 face_recognition 人脸检测的知识。

扫码观看本节视频讲解

5.3.1 检测人脸眼睛的状态

请看下面的实例文件 blink_detection.py，可以从相机中检测眼睛的状态。如果用户的眼睛闭上几秒钟，系统将打印输出“眼睛闭上”，直到用户按空格键确认此状态为止。注意，本实例需要在 Linux 系统下运行，并且必须以 sudo 权限运行键盘模块才能正常工作。

源码路径：daima\5\5-10\blink_detection.py

```
import face_recognition
import cv2
import time
from scipy.spatial import distance as dist

EYES_CLOSED_SECONDS = 5

def main():
    closed_count = 0
    video_capture = cv2.VideoCapture(0)

    ret, frame = video_capture.read(0)
    small_frame = cv2.resize(frame, (0, 0), fx=0.25, fy=0.25)
    rgb_small_frame = small_frame[:, :, ::-1]

    face_landmarks_list = face_recognition.face_landmarks(rgb_small_frame)
    process = True

    while True:
        ret, frame = video_capture.read(0)

        # 转换成正确的格式
        small_frame = cv2.resize(frame, (0, 0), fx=0.25, fy=0.25)
        rgb_small_frame = small_frame[:, :, ::-1]
        # 获得正确的面部标志

        if process:
            face_landmarks_list = face_recognition.face_landmarks(rgb_small_frame)
```

```
        #抓住眼睛
        for face_landmark in face_landmarks_list:
            left_eye = face_landmark['left_eye']
            right_eye = face_landmark['right_eye']
            color = (255,0,0)
            thickness = 2
            cv2.rectangle(small_frame, left_eye[0], right_eye[-1], color,
                          thickness)
            cv2.imshow('Video', small_frame)

            ear_left = get_ear(left_eye)
            ear_right = get_ear(right_eye)
            closed = ear_left < 0.2 and ear_right < 0.2
            if (closed):
                closed_count += 1

            else:
                closed_count = 0
            if (closed_count >= EYES_CLOSED_SECONDS):
                asleep = True
                while (asleep): #继续此循环，直到他们醒来并确认音乐
                    print("眼睛闭上")

                    if cv2.waitKey(1) == 32: #等待空格键
                        asleep = False
                        print("眼睛打开")
                closed_count = 0

    process = not process
    key = cv2.waitKey(1) & 0xFF
    if key == ord("q"):
        break

def get_ear(eye):

    #计算两个(x, y)坐标之间的欧氏距离
    A = dist.euclidean(eye[1], eye[5])
    B = dist.euclidean(eye[2], eye[4])

    # 计算(x, y)坐标之间的水平欧氏距离
    C = dist.euclidean(eye[0], eye[3])

    #计算眼睛纵横比
    ear = (A + B) / (2.0 * C)
    # 返回眼睛纵横比
    return ear

if __name__ == "__main__":
    main()
```

执行代码后的效果如图 5-9 所示。

图 5-9　执行效果

5.3.2　模糊处理人脸

在现实应用中，有时候需要保护个人的隐私，如在电视节目中将人脸进行马赛克处理。请看下面的实例文件 blur_faces_on_webcam.py，功能是使用 OpenCV 读取摄像头中的人脸数据，然后将检测到的人脸实现模糊处理。

源码路径：daima\5\5-11\blur_faces_on_webcam.py

```
import face_recognition
import cv2

#获取对摄像头 0 的引用(0 是默认值)
video_capture = cv2.VideoCapture(0)

# 初始化一些变量
face_locations = []

while True:
    #抓拍了一帧视频
    ret, frame = video_capture.read()
    # 将视频帧的大小调整为 1/4，以便更快地进行人脸检测处理
    small_frame = cv2.resize(frame, (0, 0), fx=0.25, fy=0.25)
    #查找当前视频帧中的所有面和面编码
    face_locations = face_recognition.face_locations(small_frame, model="cnn")

    #显示结果
    for top, right, bottom, left in face_locations:
        # 缩放面部位置，检测到的帧已缩放为 1/4 大小
        top *= 4
        right *= 4
```

```
        bottom *= 4
        left *= 4

        # 提取包含人脸的图像区域
        face_image = frame[top:bottom, left:right]

        # 模糊面部图像
        face_image = cv2.GaussianBlur(face_image, (99, 99), 30)

        # 将模糊的人脸区域放回帧图像中
        frame[top:bottom, left:right] = face_image

    #显示结果图像
    cv2.imshow('Video', frame)

    # 按键盘中的q键退出
    if cv2.waitKey(1) & 0xFF == ord('q'):
        break

#释放摄像头资源
video_capture.release()
cv2.destroyAllWindows()
```

执行代码后会模糊处理摄像头中的人脸，效果如图 5-10 所示。

图 5-10 执行效果

5.3.3 检测两个人脸是否匹配

在现实应用中检查两张脸是否匹配(真或假)，通常是通过验证相似度实现的。在库 face_recognition 中，通过内置函数 face_distance()来比较两张脸的相似度。函数 face_distance()通过提供一组面部编码，将它们与已知的面部编码进行比较，得到欧氏距离。对于每一个比较的脸来说，欧氏距离代表了这些脸有多相似。函数 face_distance()语法格式如下：

```
face_distance(face_encodings, face_to_compare)
```

参数说明如下。

- faces：要比较的人脸编码列表。
- face_to_compare：待进行对比的单张人脸编码数据。
- tolerance：两张脸之间有多少距离才算匹配。该值越小对比越严格，0.6 是典型的最佳值。
- 返回值：一个 NumPy ndarray，数组中的欧式距离与 faces 数组的顺序一一对应。

请看下面的实例文件 face_distance.py，功能是使用函数 face_distance()检测两张人脸是否匹配。本实例模型的训练方式是：距离小于等于 0.6 的脸是匹配的。但如果读者朋友们想更加严格，可以设置一个较小的脸距离。例如，使用数值 0.55 会减少假阳性匹配，但同时会有更多假阴性的风险。

源码路径：daima\5\5-12\face_distance.py

```
import face_recognition

#加载两幅图像进行比较
known_obama_image = face_recognition.load_image_file("obama.jpg")
known_biden_image = face_recognition.load_image_file("biden.jpg")

#获取已知图像的人脸编码
obama_face_encoding = face_recognition.face_encodings(known_obama_image)[0]
biden_face_encoding = face_recognition.face_encodings(known_biden_image)[0]

known_encodings = [
    obama_face_encoding,
    biden_face_encoding
]

#加载一个测试图像并获取它的编码
image_to_test = face_recognition.load_image_file("obama2.jpg")
image_to_test_encoding = face_recognition.face_encodings(image_to_test)[0]

# 查看测试图像与已知面之间的距离
face_distances = face_recognition.face_distance(known_encodings,
image_to_test_encoding)

for i, face_distance in enumerate(face_distances):
    print("The test image has a distance of {:.2} from known image
        #{}".format(face_distance, i))
    print("- With a normal cutoff of 0.6, would the test image match the known
        image? {}".format(face_distance < 0.6))
```

```
print("- With a very strict cutoff of 0.5, would the test image match the
      known image? {}".format(face_distance < 0.5))
print()
```

执行代码后会比较两幅照片 obama.jpg 和 biden.jpg 的人脸相似度，输出：

```
The test image has a distance of 0.35 from known image #0
- With a normal cutoff of 0.6, would the test image match the known image? True
- With a very strict cutoff of 0.5, would the test image match the known image? True

The test image has a distance of 0.82 from known image #1
- With a normal cutoff of 0.6, would the test image match the known image? False
- With a very strict cutoff of 0.5, would the test image match the known image? False
```

5.3.4 识别视频中的人脸

请看下面的实例文件 facerec_from_video_file.py，功能是识别某个视频文件中的人脸，然后将结果保存到新的视频文件中。

源码路径：daima\5\5-13\facerec_from_video_file.py

```
import face_recognition
import cv2

#打开要识别的视频文件
input_movie = cv2.VideoCapture("hamilton_clip.mp4")
length = int(input_movie.get(cv2.CAP_PROP_FRAME_COUNT))

#创建输出电影文件(确保分辨率/帧速率与输入视频匹配)
fourcc = cv2.VideoWriter_fourcc(*'XVID')
output_movie = cv2.VideoWriter('output.avi', fourcc, 29.97, (640, 360))

#加载实例图片，并学习如何识别它们
lmm_image = face_recognition.load_image_file("lin-manuel-miranda.png")
lmm_face_encoding = face_recognition.face_encodings(lmm_image)[0]

al_image = face_recognition.load_image_file("alex-lacamoire.png")
al_face_encoding = face_recognition.face_encodings(al_image)[0]

known_faces = [
   lmm_face_encoding,
   al_face_encoding
]

# 初始化一些变量
face_locations = []
face_encodings = []
```

```
face_names = []
frame_number = 0

while True:
    #抓取一帧视频
    ret, frame = input_movie.read()
    frame_number += 1

    #输入视频文件结束时退出
    if not ret:
        break

    # 将图像从BGR颜色(OpenCV使用)转换为RGB颜色(人脸识别使用的颜色)
    rgb_frame = frame[:, :, ::-1]

    #查找当前视频帧中的所有面和面编码
    face_locations = face_recognition.face_locations(rgb_frame)
    face_encodings = face_recognition.face_encodings(rgb_frame, face_locations)

    face_names = []
    for face_encoding in face_encodings:
        #查看该面是否与已知面匹配
        match = face_recognition.compare_faces(known_faces, face_encoding,
                                               tolerance=0.50)

        #如果有两张以上的识别脸，则可以使当前编码逻辑变得更漂亮，但我保持简单的演示
        name = None
        if match[0]:
            name = "Lin-Manuel Miranda"
        elif match[1]:
            name = "Alex Lacamoire"

        face_names.append(name)

    #标记结果
    for (top, right, bottom, left), name in zip(face_locations, face_names):
        if not name:
            continue

        #在脸上画一个方框
        cv2.rectangle(frame, (left, top), (right, bottom), (0, 0, 255), 2)

        #在脸的下方绘制一个带有名称的标签
        cv2.rectangle(frame, (left, bottom - 25), (right, bottom), (0, 0, 255),
                      cv2.FILLED)
        font = cv2.FONT_HERSHEY_DUPLEX
        cv2.putText(frame, name, (left + 6, bottom - 6), font, 0.5, (255, 255,
                    255), 1)
```

```
    #将识别结果图像写入到输出视频文件中
    print("Writing frame {} / {}".format(frame_number, length))
    output_movie.write(frame)

input_movie.release()
cv2.destroyAllWindows()
```

上述代码的具体说明如下。

(1) 首先准备了视频文件 hamilton_clip.mp4 作为输入文件，然后设置输出文件名为 output.avi。

(2) 准备素材图片文件 lin-manuel-miranda.png，在此文件中保存的是一幅人脸照片。

(3) 处理输入视频文件 hamilton_clip.mp4，在视频中标记处图片文件 lin-manuel-miranda.png 中的人脸，并将检测结果保存为输出视频文件 output.avi。

执行代码后会检测输入视频文件 hamilton_clip.mp4 中的每一帧，并标记出图片文件 lin-manuel-miranda.png 中的人脸。打开识别结果视频文件 output.avi，效果如图 5-11 所示。

图 5-11 执行效果

5.3.5 网页版人脸识别器

请看下面的实例文件 web_service_example.py，它基于 Flask 框架开发的一个在线 Web 程序。在 Web 网页中可以上传图片到服务器，然后识别这幅上传图片中的人脸是不是奥巴马，并使用 JSON 键值对输出显示识别结果。

源码路径：daima\5\5-14\web_service_example.py

```
import face_recognition
from flask import Flask, jsonify, request, redirect

#可以将其更改为系统上的任何文件夹
ALLOWED_EXTENSIONS = {'png', 'jpg', 'jpeg', 'gif'}
```

```
app = Flask(__name__)


def allowed_file(filename):
    return '.' in filename and \
           filename.rsplit('.', 1)[1].lower() in ALLOWED_EXTENSIONS

@app.route('/', methods=['GET', 'POST'])
def upload_image():
    # 检测图片是否上传成功
    if request.method == 'POST':
        if 'file' not in request.files:
            return redirect(request.url)

        file = request.files['file']

        if file.filename == '':
            return redirect(request.url)

        if file and allowed_file(file.filename):
            # 图片上传成功，检测图片中的人脸
            return detect_faces_in_image(file)

    # 图片上传失败，输出以下html代码
    return '''
    <!doctype html>
    <title>Is this a picture of Obama?</title>
    <h1>Upload a picture and see if it's a picture of Obama!</h1>
    <form method="POST" enctype="multipart/form-data">
      <input type="file" name="file">
      <input type="submit" value="Upload">
    </form>
    '''

def detect_faces_in_image(file_stream):
    # 用face_recognition.face_encodings(img)接口提前把奥巴马人脸的编码录入
    known_face_encoding = [-0.09634063,  0.12095481, -0.00436332, -0.07643753,
                            0.0080383,
                         0.01902981, -0.07184699, -0.09383309,  0.18518871,
                            -0.09588896,
                         0.23951106,  0.0986533 , -0.22114635, -0.1363683 ,
                            0.04405268,
                         0.11574756, -0.19899382, -0.09597053, -0.11969153,
                            -0.12277931,
                         0.03416885, -0.00267565,  0.09203379,  0.04713435,
                            -0.12731361,
```

```
          -0.35371891, -0.0503444 , -0.17841317, -0.00310897,
              -0.09844551,
          -0.06910533, -0.00503746, -0.18466514, -0.09851682,
              0.02903969,
          -0.02174894,  0.02261871,  0.0032102 ,  0.20312519,
              0.02999607,
          -0.11646006,  0.09432904,  0.02774341,  0.22102901,
              0.26725179,
           0.06896867, -0.00490024, -0.09441824,  0.11115381,
              -0.22592428,
           0.06230862,  0.16559327,  0.06232892,  0.03458837,
              0.09459756,
          -0.18777156,  0.00654241,  0.08582542, -0.13578284,
              0.0150229,
           0.00670836, -0.08195844, -0.04346499,  0.03347827,
              0.20310158,
           0.09987706, -0.12370517, -0.06683611,  0.12704916,
              -0.02160804,
           0.00984683,  0.00766284, -0.18980607, -0.19641446,
              -0.22800779,
           0.09010898,  0.39178532,  0.18818057, -0.20875394,
              0.03097027,
          -0.21300618,  0.02532415,  0.07938635,  0.01000703,
              -0.07719778,
          -0.12651891, -0.04318593,  0.06219772,  0.09163868,
              0.05039065,
          -0.04922386,  0.21839413, -0.02394437,  0.06173781,
              0.0292527,
           0.06160797, -0.15553983, -0.02440624, -0.17509389,
              -0.0630486,
           0.01428208, -0.03637431,  0.03971229,  0.13983178,
              -0.23006812,
           0.04999552,  0.0108454 , -0.03970895,  0.02501768,
              0.08157793,
          -0.03224047, -0.04502571,  0.0556995 , -0.24374914,
              0.25514284,
           0.24795187,  0.04060191,  0.17597422,  0.07966681,
              0.01920104,
          -0.01194376, -0.02300822, -0.17204897, -0.0596558 ,
              0.05307484,
           0.07417042,  0.07126575,  0.00209804]

# 载入用户上传的图片
img = face_recognition.load_image_file(file_stream)
# 为用户上传的图片中的人脸编码
unknown_face_encodings = face_recognition.face_encodings(img)
```

```
    face_found = False
    is_obama = False

    if len(unknown_face_encodings) > 0:
        face_found = True
        # 看看图片中的第一张脸是不是奥巴马
        match_results = face_recognition.compare_faces([known_face_encoding],
                                                       unknown_face_encodings[0])
        if match_results[0]:
            is_obama = True

    # 将识别结果以 JSON 键值对的数据结构输出
    result = {
        "face_found_in_image": face_found,
        "is_picture_of_obama": is_obama
    }
    return jsonify(result)


if __name__ == "__main__":
    app.run(debug=True)
```

运行上述 Flask 程序，然后在浏览器中输入 URL 地址 http://127.0.0.1:5000/，如图 5-12 所示。

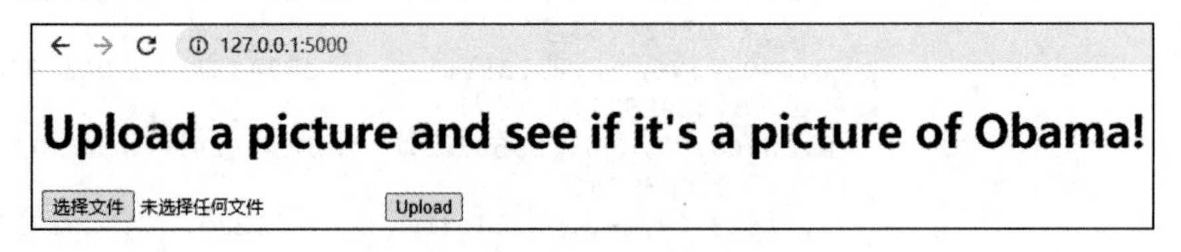

图 5-12　Flask 主页

单击“选择文件”按钮选择一幅照片，单击 Upload 按钮上传被选择的照片。然后调用 face_recognition 识别上传照片中的人物是不是奥巴马，例如上传一幅照片后会输出如图 5-13 所示的识别结果。

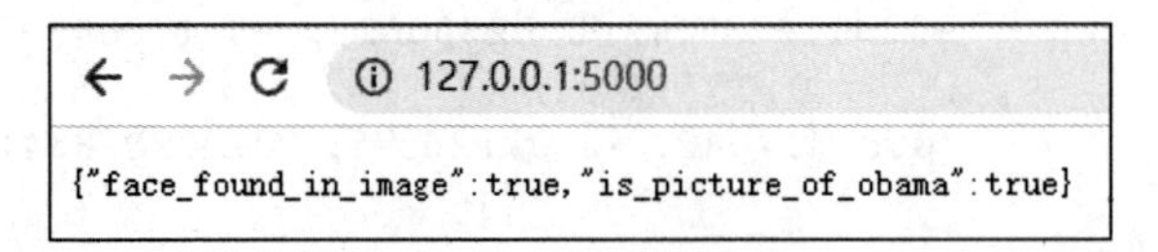

图 5-13　识别结果

第6章

Scikit-Learn 机器学习和人脸识别

Scikit-Learn(有时被简称为 Sklearn)是一个著名的 Python 机器学习库，提供了各种分类、回归和聚类算法，包括支持向量机、随机森林、梯度提升、k 均值和 DBSCAN。在本章的内容中，将详细讲解使用 Scikit-Learn 机器学习技术实现人脸识别的知识。

6.1 Scikit-Learn 基础

自从 2007 年发布以来，Scikit-Learn 已经成为重要的 Python 机器学习库之一。Scikit-Learn 是 Scipy 的扩展，建立在库 NumPy 和库 Matplolib 的基础上。利用这几大模块的优势，可以大大地提高机器学习的效率。

扫码观看本节视频讲解

6.1.1 Scikit-Learn 介绍

Scikit-Learn 拥有着完善的技术文档，容易上手；具有丰富的 API，在学术界颇受欢迎。Scikit-Learn 已经封装了大量的机器学习算法，包括 LIBSVM 和 LIBINEAR。同时内置了大量的数据集，节省了获取和整理数据集的时间。

在库 Scikit-Learn 中内置了多种算法，具体说明如下。

- 回归：线性、决策树、SVM、KNN。
- 集成回归：随机森林、Adaboost、GradientBoosting、Bagging、ExtraTrees。
- 分类：线性、决策树、SVM、KNN、朴素贝叶斯。
- 集成分类：随机森林、Adaboost、GradientBoosting、Bagging、ExtraTrees。
- 常用聚类：k 均值(K-means)、层次聚类(Hierarchical clustering)、DBSCAN。
- 常用降维：LinearDiscriminantAnalysis、PCA。

6.1.2 安装 Scikit-Learn

在安装 Scikit-Learn 之前需要先安装 Python、NumPy、Scipy 和 Matplotlib(可选)，在 Windows 系统中，可以使用如下命令安装 Scikit-Learn：

```
pip install scikit_learn
```

如果使用 pip 命令在线安装失败，可以考虑下载 whl 文件到本地，然后再使用 pip 命令安装 whl 文件的方式来安装 Scikit-Learn。

6.2 基于 Scikit-Learn 的常用算法

库 Scikit-Learn 的主要功能是实现机器学习，在本节的内容中，将详细讲解使用

Scikit-Learn 实现机器学习常用算法的知识。

6.2.1 Scikit-Learn 机器学习的基本流程

扫码观看本节视频讲解

机器学习从开始到建模的基本流程是：获取数据、数据预处理、训练模型、模型评估、预测、分类。本次我们将根据传统机器学习的流程，看看在每一步流程中都有哪些常用的函数以及它们的用法是怎么样的。请看下面的实例文件 hua.py，其实现了鸢尾花识别的功能。这是一个经典的机器学习分类问题，它的数据样本中包括了 4 个特征变量、1 个类别变量，样本总数为 150。本实例的目标是根据花萼长度(sepal length)、花萼宽度(sepal width)、花瓣长度(petal length)、花瓣宽度(petal width)这 4 个特征来识别出鸢尾花属于山鸢尾(iris-setosa)、变色鸢尾(iris-versicolor)和维吉尼亚鸢尾(iris-virginica)中的哪一种。

源码路径：daima\6\6-2\hua.py

```
# 引入数据集，sklearn 包含众多数据集
from sklearn import datasets
# 将数据分为测试集和训练集
from sklearn.model_selection import train_test_split
# 利用邻近点方式训练数据
from sklearn.neighbors import KNeighborsClassifier

# 引入数据,本次导入鸢尾花数据，iris 数据包含 4 个特征变量
iris = datasets.load_iris()
# 特征变量
iris_X = iris.data
# print(iris_X)
print('特征变量的长度', len(iris_X))
# 目标值
iris_y = iris.target
print('鸢尾花的目标值', iris_y)
# 利用 train_test_split 将训练集和测试集分开，test_size 占 30%
X_train, X_test, y_train, y_test = train_test_split(iris_X, iris_y,
test_size=0.3)
# 可以看到训练数据的特征值分为 3 类
print(y_train)

# 训练数据
# 引入训练方法
knn = KNeighborsClassifier()
# 进行填充测试
```

```
knn.fit(X_train, y_train)

params = knn.get_params()
print(params)
score = knn.score(X_test, y_test)
print("预测得分为：%s" % score)

# 预测数据，预测特征值
print(knn.predict(X_test))

# 打印真实特征值
print(y_test)
```

执行代码后会输出训练和预测结果：

```
特征变量的长度 150
鸢尾花的目标值 [0 0 0 0 0 0 0 0 0 0 0 0 0 0 0 0 0 0 0 0 0 0 0 0 0 0 0 0 0 0 0 0
0 0 0 0 0
 0 0 0 0 0 0 0 0 0 0 0 0 0 1 1 1 1 1 1 1 1 1 1 1 1 1 1 1 1 1 1 1 1 1 1 1 1
 1 1 1 1 1 1 1 1 1 1 1 1 1 1 1 1 1 1 1 1 1 1 1 1 1 1 2 2 2 2 2 2 2 2 2 2 2
 2 2 2 2 2 2 2 2 2 2 2 2 2 2 2 2 2 2 2 2 2 2 2 2 2 2 2 2 2 2 2 2 2 2 2 2 2
 2 2]
[2 1 2 1 0 2 0 1 0 1 0 1 1 2 1 0 0 0 1 2 2 2 2 1 1 2 1 0 2 0 2 0 0 2 2 2 0
 1 1 2 0 2 1 1 1 2 0 0 1 1 1 1 1 0 1 0 2 2 2 1 1 0 0 2 0 2 1 0 2 1 1 0 2 2
 2 0 1 1 0 2 0 1 2 2 1 1 1 0 1 1 2 0 0 2 0 0 1 2 0 0 0 0 1 2 2]
{'algorithm': 'auto', 'leaf_size': 30, 'metric': 'minkowski', 'metric_params':
None, 'n_jobs': None, 'n_neighbors': 5, 'p': 2, 'weights': 'uniform'}
预测得分为：1.0
[0 2 0 0 1 0 1 1 0 0 2 2 1 0 2 2 1 0 0 2 0 2 1 0 2 1 2 2 2 2 0 2 0 0 1 2 2
 0 1 2 1 1 1 0 1]
[0 2 0 0 1 0 1 1 0 0 2 2 1 0 2 2 1 0 0 2 0 2 1 0 2 1 2 2 2 2 0 2 0 0 1 2 2
 0 1 2 1 1 1 0 1]
```

6.2.2 分类算法

请看下面的实例文件 fen.py，功能是绘制不同分类器的分类概率。我们使用一个 3 类的数据集，并使用支持向量分类器、带 L1 和 L2 惩罚项的 Logistic 回归， 使用 One-Vs-Rest 或多项设置以及高斯过程分类对其进行分类。在默认情况下，线性 SVC 不是概率分类器，但在本例中它有一个内建校准选项(probability=True)。箱外的 One-Vs-Rest 的逻辑回归不是一个多分类的分类器，因此，与其他估计器相比，它在分离第 2 类和第 3 类时有更大的困难。

源码路径：daima\6\6-2\fen.py

```
import matplotlib.pyplot as plt
import numpy as np

from sklearn.metrics import accuracy_score
from sklearn.linear_model import LogisticRegression
from sklearn.svm import SVC
from sklearn.gaussian_process import GaussianProcessClassifier
from sklearn.gaussian_process.kernels import RBF
from sklearn import datasets

iris = datasets.load_iris()
X = iris.data[:, 0:2]  # we only take the first two features for visualization
y = iris.target

n_features = X.shape[1]

C = 10
kernel = 1.0 * RBF([1.0, 1.0])  # for GPC

# Create different classifiers.
classifiers = {
    'L1 logistic': LogisticRegression(C=C, penalty='l1',
                                      solver='saga',
                                      multi_class='multinomial',
                                      max_iter=10000),
    'L2 logistic (Multinomial)': LogisticRegression(C=C, penalty='l2',
                                                    solver='saga',
                                                    multi_class='multinomial',
                                                    max_iter=10000),
    'L2 logistic (OvR)': LogisticRegression(C=C, penalty='l2',
                                            solver='saga',
                                            multi_class='ovr',
                                            max_iter=10000),
    'Linear SVC': SVC(kernel='linear', C=C, probability=True,
                      random_state=0),
    'GPC': GaussianProcessClassifier(kernel)
}

n_classifiers = len(classifiers)

plt.figure(figsize=(3 * 2, n_classifiers * 2))
plt.subplots_adjust(bottom=.2, top=.95)
```

```
xx = np.linspace(3, 9, 100)
yy = np.linspace(1, 5, 100).T
xx, yy = np.meshgrid(xx, yy)
Xfull = np.c_[xx.ravel(), yy.ravel()]

for index, (name, classifier) in enumerate(classifiers.items()):
    classifier.fit(X, y)

    y_pred = classifier.predict(X)
    accuracy = accuracy_score(y, y_pred)
    print("Accuracy (train) for %s: %0.1f%% " % (name, accuracy * 100))

    # View probabilities:
    probas = classifier.predict_proba(Xfull)
    n_classes = np.unique(y_pred).size
    for k in range(n_classes):
        plt.subplot(n_classifiers, n_classes, index * n_classes + k + 1)
        plt.title("Class %d" % k)
        if k == 0:
            plt.ylabel(name)
        imshow_handle = plt.imshow(probas[:, k].reshape((100, 100)),
                                   extent=(3, 9, 1, 5), origin='lower')
        plt.xticks(())
        plt.yticks(())
        idx = (y_pred == k)
        if idx.any():
            plt.scatter(X[idx, 0], X[idx, 1], marker='o', c='w', edgecolor='k')

ax = plt.axes([0.15, 0.04, 0.7, 0.05])
plt.title("Probability")
plt.colorbar(imshow_handle, cax=ax, orientation='horizontal')

plt.show()
```

执行代码后会输出下面的结果，并在 Matplotlib 中绘制 3 种分类的概率，如图 6-1 所示。

```
Accuracy (train) for L1 logistic: 83.3%
Accuracy (train) for L2 logistic (Multinomial): 82.7%
Accuracy (train) for L2 logistic (OvR): 79.3%
Accuracy (train) for Linear SVC: 82.0%
Accuracy (train) for GPC: 82.7%
```

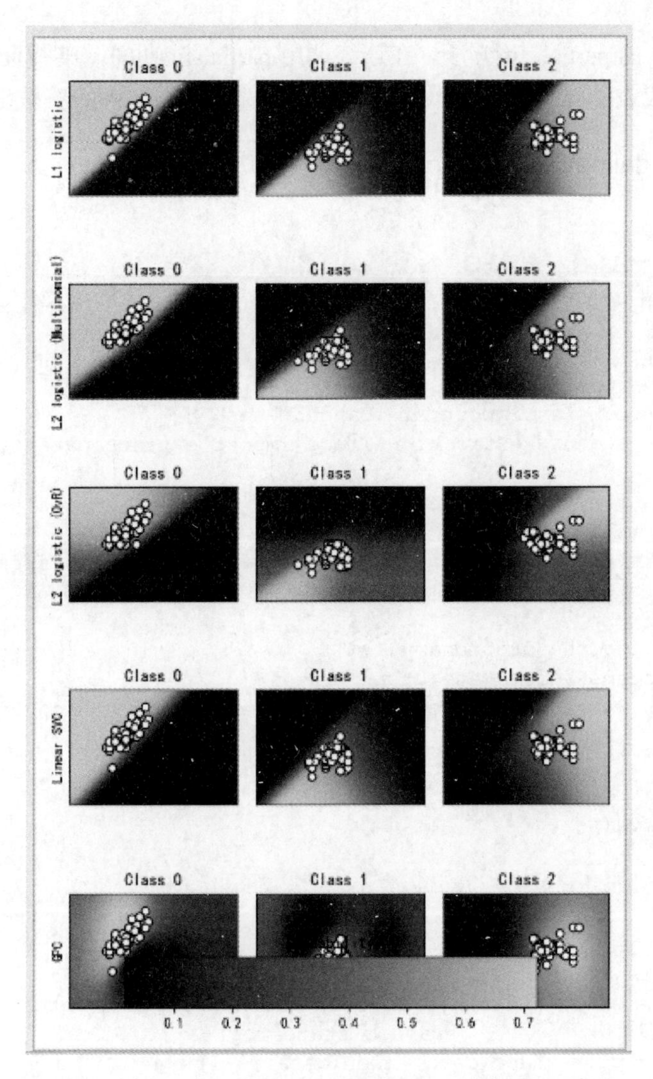

图 6-1　执行效果

6.2.3　聚类算法

请看下面的实例文件 face.py，功能是使用一个大型的 Faces 数据集学习一组组成面部 20×20 的图像修补程序。本实例非常有趣，展示了使用 Scikit-Learn 在线 API 学习按块处理一个大型数据集的方法。本实例处理的方法是一次加载一个图像，并从这个图像中随机提取 50 个补丁。一旦积累了 500 个补丁(使用 10 个图像)，则运行在线 KMeans 对象

MiniBatchKMeans 的 partial_fit 方法。在连续调用 partial-fit 期间，某些聚类会被重新分配，这是因为它们所代表的补丁数量太少了，所以最好选择一个随机的新聚类。

源码路径：daima\6\6-2\face.py

```
import time

import matplotlib.pyplot as plt
import numpy as np


from sklearn import datasets
from sklearn.cluster import MiniBatchKMeans
from sklearn.feature_extraction.image import extract_patches_2d

faces = datasets.fetch_olivetti_faces()

# ############################################################################
# Learn the dictionary of images

print('Learning the dictionary... ')
rng = np.random.RandomState(0)
kmeans = MiniBatchKMeans(n_clusters=81, random_state=rng, verbose=True)
patch_size = (20, 20)

buffer = []
t0 = time.time()

# 在整个数据集上循环 6 次
index = 0
for _ in range(6):
    for img in faces.images:
        data = extract_patches_2d(img, patch_size, max_patches=50,
                                  random_state=rng)
        data = np.reshape(data, (len(data), -1))
        buffer.append(data)
        index += 1
        if index % 10 == 0:
            data = np.concatenate(buffer, axis=0)
            data -= np.mean(data, axis=0)
            data /= np.std(data, axis=0)
            kmeans.partial_fit(data)
            buffer = []
        if index % 100 == 0:
            print('Partial fit of %4i out of %i'
                  % (index, 6 * len(faces.images)))
```

```
dt = time.time() - t0
print('done in %.2fs.' % dt)

# ############################################################################
# Plot the results
plt.figure(figsize=(4.2, 4))
for i, patch in enumerate(kmeans.cluster_centers_):
   plt.subplot(9, 9, i + 1)
   plt.imshow(patch.reshape(patch_size), cmap=plt.cm.gray,
           interpolation='nearest')
   plt.xticks(())
   plt.yticks(())


plt.suptitle('Patches of faces\nTrain time %.1fs on %d patches' %
          (dt, 8 * len(faces.images)), fontsize=16)
plt.subplots_adjust(0.08, 0.02, 0.92, 0.85, 0.08, 0.23)

plt.show()
```

执行效果如图 6-2 所示。

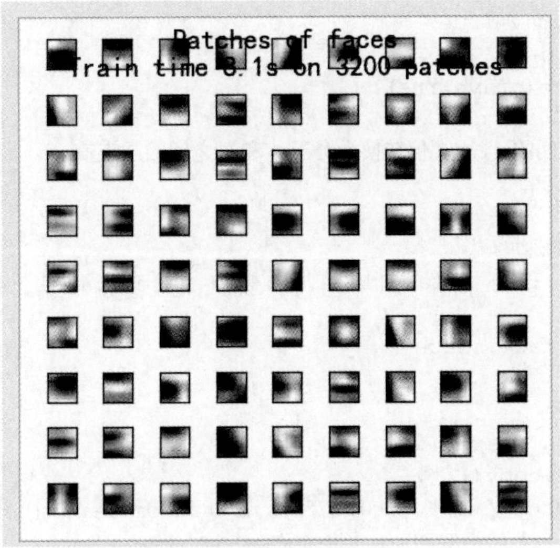

图 6-2 执行效果

6.2.4 分解算法

请看下面的实例文件 zidian.py，演示了基于字典学习的图像去噪的应用过程。在本实例中处理了一幅浣熊照片，比较了浣熊人脸图像噪声碎片重构效果，用到了在线字典学习和各种转换方法。字典用来拟合浣熊图像的左半部，然后用来重建浣熊的右半部分。请注意：更好的性能可以通过拟合一个不失真(即无噪音)图像来实现，但在这里我们假设它是不可用的。

源码路径：daima\6\6-2\zidian.py

```
from time import time

import matplotlib.pyplot as plt
import numpy as np
import scipy as sp

from sklearn.decomposition import MiniBatchDictionaryLearning
from sklearn.feature_extraction.image import extract_patches_2d
from sklearn.feature_extraction.image import reconstruct_from_patches_2d


try:  # SciPy >= 0.16 在misc中有 face
    from scipy.misc import face
    face = face(gray=True)
except ImportError:
    face = sp.face(gray=True)

# 将介于0和255之间的uint8值转换为介于0和1之间的浮点值
face = face / 255.

# 低采样以获得更高的速度
face = face[::4, ::4] + face[1::4, ::4] + face[::4, 1::4] + face[1::4, 1::4]
face /= 4.0
height, width = face.shape

# 扭曲图像的右半部分
print('Distorting image...')
distorted = face.copy()
distorted[:, width // 2:] += 0.075 * np.random.randn(height, width // 2)

#从图像的左半部分提取所有参考patches
print('Extracting reference patches...')
t0 = time()
patch_size = (7, 7)
data = extract_patches_2d(distorted[:, :width // 2], patch_size)
data = data.reshape(data.shape[0], -1)
data -= np.mean(data, axis=0)
```

```
data /= np.std(data, axis=0)
print('done in %.2fs.' % (time() - t0))

# ###########################################################################
# 从参考补丁中学习词典

print('Learning the dictionary...')
t0 = time()
dico = MiniBatchDictionaryLearning(n_components=100, alpha=1, n_iter=500)
V = dico.fit(data).components_
dt = time() - t0
print('done in %.2fs.' % dt)

plt.figure(figsize=(4.2, 4))
for i, comp in enumerate(V[:100]):
   plt.subplot(10, 10, i + 1)
   plt.imshow(comp.reshape(patch_size), cmap=plt.cm.gray_r,
            interpolation='nearest')
   plt.xticks(())
   plt.yticks(())
plt.suptitle('Dictionary learned from face patches\n' +
          'Train time %.1fs on %d patches' % (dt, len(data)),
          fontsize=16)
plt.subplots_adjust(0.08, 0.02, 0.92, 0.85, 0.08, 0.23)


# ###########################################################################
# 显示扭曲图像

def show_with_diff(image, reference, title):
   """显示去噪的辅助函数"""
   plt.figure(figsize=(5, 3.3))
   plt.subplot(1, 2, 1)
   plt.title('Image')
   plt.imshow(image, vmin=0, vmax=1, cmap=plt.cm.gray,
            interpolation='nearest')
   plt.xticks(())
   plt.yticks(())
   plt.subplot(1, 2, 2)
   difference = image - reference

   plt.title('Difference (norm: %.2f)' % np.sqrt(np.sum(difference ** 2)))
   plt.imshow(difference, vmin=-0.5, vmax=0.5, cmap=plt.cm.PuOr,
            interpolation='nearest')
   plt.xticks(())
```

```
    plt.yticks(())
    plt.suptitle(title, size=16)
    plt.subplots_adjust(0.02, 0.02, 0.98, 0.79, 0.02, 0.2)

show_with_diff(distorted, face, 'Distorted image')

# ############################################################################
# 提取噪声斑块并利用字典进行重构

print('Extracting noisy patches... ')
t0 = time()
data = extract_patches_2d(distorted[:, width // 2:], patch_size)
data = data.reshape(data.shape[0], -1)
intercept = np.mean(data, axis=0)
data -= intercept
print('done in %.2fs.' % (time() - t0))

transform_algorithms = [
    ('Orthogonal Matching Pursuit\n1 atom', 'omp',
     {'transform_n_nonzero_coefs': 1}),
    ('Orthogonal Matching Pursuit\n2 atoms', 'omp',
     {'transform_n_nonzero_coefs': 2}),
    ('Least-angle regression\n5 atoms', 'lars',
     {'transform_n_nonzero_coefs': 5}),
    ('Thresholding\n alpha=0.1', 'threshold', {'transform_alpha': .1})]

reconstructions = {}
for title, transform_algorithm, kwargs in transform_algorithms:
    print(title + '...')
    reconstructions[title] = face.copy()
    t0 = time()
    dico.set_params(transform_algorithm=transform_algorithm, **kwargs)
    code = dico.transform(data)
    patches = np.dot(code, V)

    patches += intercept
    patches = patches.reshape(len(data), *patch_size)
    if transform_algorithm == 'threshold':
        patches -= patches.min()
        patches /= patches.max()
    reconstructions[title][:, width // 2:] = reconstruct_from_patches_2d(
        patches, (height, width // 2))
    dt = time() - t0
```

```
    print('done in %.2fs.' % dt)
    show_with_diff(reconstructions[title], face,
                   title + ' (time: %.1fs)' % dt)

plt.show()
```

评价图像去噪效果的一个常见方法是：通过观察重建图像与原始图像的差异来评价图像去噪效果。如果重建是完美的，这将看起来像高斯噪声。本实例的执行效果如图 6-3 所示，从图 6-3 中可以看出，具有两个非零系数的正交匹配追踪(OMP)的结果比只保持一个(边界看起来不那么突出)的结果有一点偏差，最小角回归的结果具有更强的偏差：这种差异使人联想到原始图像的局部强度值。

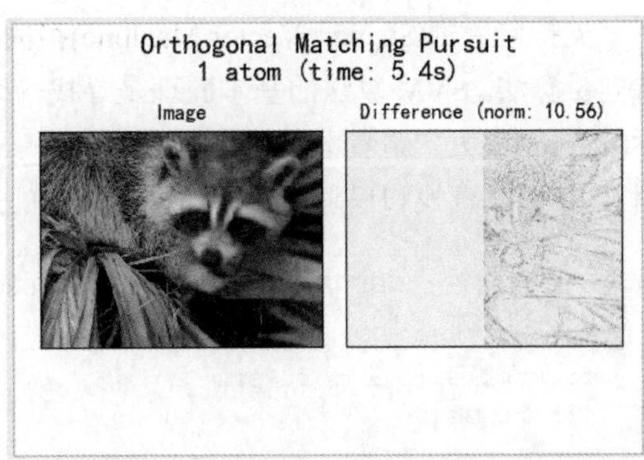

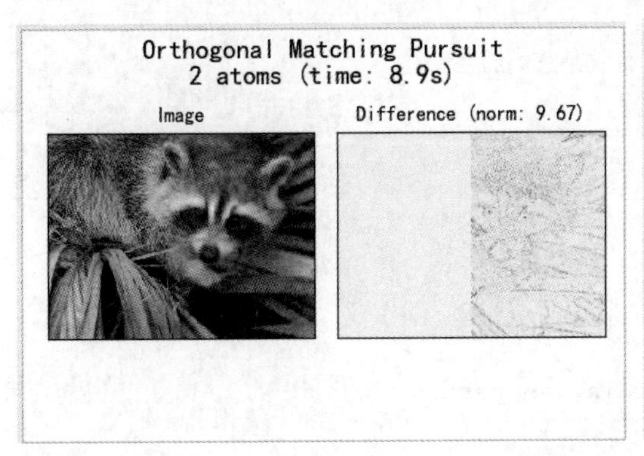

图 6-3 执行效果

从上数执行效果可以看出，阈值处理对去噪并没有帮助，但是可以表明阈值能够以非常高的速度产生暗示性的输出。因此阈值对其他任务(如目标分类)会非常有用，在这些任务中，性能不一定与可视化有关。

6.3 Scikit-Learn 和人脸识别

经过本章前面内容的学习，已经基本上了解了 Scikit-Learn 常用机器学习算法的知识。在本节的内容中，将详细讲解使用 Scikit-Learn 实现人脸识别的知识。

扫码观看本节视频讲解

6.3.1 SVM 算法人脸识别

支持向量机(Support Vector Machine)指的是一系列机器学习方法，这类方法的基础是支持向量算法。SVM 算法的基本原理是寻找一个能够区分两类的超平面(hyper plane)，使得边际(margin)最大。请看下面的实例文件 face_recognition_svm.py，其演示基于 Scikit-Learn 使用 SVC 算法查找和识别指定图像中的人脸的过程。

源码路径：daima\6\6-3\face_recognition_svm.py

```
import face_recognition
from sklearn import svm
import os

#SVC 分类器的训练
#训练数据是来自所有已知图像的所有人脸编码，标签是它们的名称
encodings = []
names = []

#训练目录
train_dir = os.listdir('knn_examples/train/')

# Loop through each person in the training directory
for person in train_dir:
    pix = os.listdir("knn_examples/train/" + person)

    #循环浏览当前人员的每个训练图像
    for person_img in pix:
        # 获取每个图像文件中的人脸编码
        face = face_recognition.load_image_file("knn_examples/train/" + person
                                                 + "/" + person_img)
        face_bounding_boxes = face_recognition.face_locations(face)

        #如果训练图像只包含一张脸
```

```
        if len(face_bounding_boxes) == 1:
            face_enc = face_recognition.face_encodings(face)[0]
            #在训练数据中为当前图像添加带有相应标签(名称)的人脸编码
            encodings.append(face_enc)
            names.append(person)
        else:
            print(person + "/" + person_img + " was skipped and can't be used for
                training")

#创建并训练 SVC 分类器
clf = svm.SVC(gamma='scale')
clf.fit(encodings,names)

#将具有未知脸的测试图像加载到 Numpy 数组中
test_image = face_recognition.load_image_file('111.jpg')

#使用默认的基于 HOG 的模型查找测试图像中的所有人脸
face_locations = face_recognition.face_locations(test_image)
no = len(face_locations)
print("Number of faces detected: ", no)

#使用训练好的分类器对测试图像中的所有人脸进行预测
print("Found:")
for i in range(no):
    test_image_enc = face_recognition.face_encodings(test_image)[i]
    name = clf.predict([test_image_enc])
    print(*name)
```

执行代码后会输出显示识别某照片中人脸的结果：

```
Number of faces detected:  1
Found:
biden
```

6.3.2 KNN 算法人脸识别

请看下面的实例文件 face_recognition_knn.py，这是一个使用 K 最近邻(k-Nearest Neighbor，KNN)分类算法实现人脸识别的例子。本实例可以识别出多个已知的人脸，并在一个可行的计算时间内对未知人进行预测。具体流程如下。

(1) 准备一组想认识的人的照片，然后在单个目录中组织图像，并为每个人设置一个子目录。

(2) 使用适当的参数运行训练函数 train()，将模型保存到本地磁盘，在下次使用时无须重新训练即可重新使用模型。

(3) 通过传递过训练的模型调用预测函数 show_prediction_labels_on_image()，识别出未知图像中的人。

源码路径：daima\6\6-3\face_recognition_knn.py

```
import math
from sklearn import neighbors
import os
import os.path
import pickle
from PIL import Image, ImageDraw
import face_recognition
from face_recognition.face_recognition_cli import image_files_in_folder

ALLOWED_EXTENSIONS = {'png', 'jpg', 'jpeg'}


def train(train_dir, model_save_path=None, n_neighbors=None,
knn_algo='ball_tree', verbose=False):
    """
  创建训练函数 train()：使用 k 近邻分类器进行人脸识别

: param train_dir：包含每个已知人员的子目录及其名称的目录。

    X = []
    y = []

    #对训练集中的每个人进行循环
    for class_dir in os.listdir(train_dir):
        if not os.path.isdir(os.path.join(train_dir, class_dir)):
            continue

        #循环浏览当前人员的每个训练图像
        for img_path in image_files_in_folder(os.path.join(train_dir, class_dir)):
            image = face_recognition.load_image_file(img_path)
            face_bounding_boxes = face_recognition.face_locations(image)

            if len(face_bounding_boxes) != 1:
                #如果训练图像中没有人(或人太多)，请跳过该图像
                if verbose:
                    print("Image {} not suitable for training: {}".format(img_path,
                            "Didn't find a face" if len(face_bounding_boxes) < 1 else
                            "Found more than one face"))
```

```
            else:
                #将当前图像的人脸编码添加到训练集中
                X.append(face_recognition.face_encodings(image, known_face_
                         locations=face_bounding_boxes)[0])
                y.append(class_dir)

    # 确定在 KNN 分类器中用于加权的近邻数
    if n_neighbors is None:
        n_neighbors = int(round(math.sqrt(len(X))))
        if verbose:
            print("Chose n_neighbors automatically:", n_neighbors)

    # 创建并训练 KNN 分类器
    knn_clf = neighbors.KNeighborsClassifier(n_neighbors=n_neighbors,
                                    algorithm=knn_algo, weights='distance')
    knn_clf.fit(X, y)

    #保存经过训练的 KNN 分类器
    if model_save_path is not None:
        with open(model_save_path, 'wb') as f:
            pickle.dump(knn_clf, f)

    return knn_clf


def predict(X_img_path, knn_clf=None, model_path=None, distance_threshold=0.6):
    """
   使用训练好的 KNN 分类器识别给定图像中的人脸

: param X_img\u path：要识别的图像的路径

: param knn_clf：(可选)knn 分类器对象。如果未指定，则必须指定型号\保存\路径。

: param model_path：(可选)到 pickled knn 分类器的路径。如果未指定，则模型\保存\路径必
须为 knn\ clf。

: param distance_threshold：(可选)人脸分类的距离阈值。它越大，就会把一个不认识的人误分
类为一个已知的人。

: return:图像中已识别面的名称和面位置列表：[(名称，边界框)，…]。

对于未识别的人脸，将返回名称“未知”。
    """
    if not os.path.isfile(X_img_path) or os.path.splitext(X_img_path)[1][1:]
                   not in ALLOWED_EXTENSIONS:
```

```
        raise Exception("Invalid image path: {}".format(X_img_path))

    if knn_clf is None and model_path is None:
        raise Exception("Must supply knn classifier either thourgh knn_clf or
                        model_path")

    # 加载经过训练的 KNN 模型(如果传入)
    if knn_clf is None:
        with open(model_path, 'rb') as f:
            knn_clf = pickle.load(f)

    # 加载图像文件并查找人脸位置
    X_img = face_recognition.load_image_file(X_img_path)
    X_face_locations = face_recognition.face_locations(X_img)

    #如果在图像中找不到脸部，则返回空结果。
    if len(X_face_locations) == 0:
        return []

    #在测试图像中查找人脸编码
    faces_encodings = face_recognition.face_encodings(X_img,
                                known_face_locations=X_face_locations)

    #使用 KNN 模型找到测试人脸的最佳匹配
    closest_distances = knn_clf.kneighbors(faces_encodings, n_neighbors=1)
    are_matches = [closest_distances[0][i][0] <= distance_threshold for i in
                  range(len(X_face_locations))]

    # 预测类并删除不在阈值内的分类
    return [(pred, loc) if rec else ("unknown", loc) for pred, loc, rec in
           zip(knn_clf.predict(faces_encodings), X_face_locations, are_matches)]


def show_prediction_labels_on_image(img_path, predictions):
    """
    显示人脸识别结果
    : param img\u path: 要识别的图像的路径
    : param predictions: 预测函数的结果
    """
    pil_image = Image.open(img_path).convert("RGB")
    draw = ImageDraw.Draw(pil_image)

    for name, (top, right, bottom, left) in predictions:
        #使用 Pillow 模块在脸部周围画一个方框
```

```
        draw.rectangle(((left, top), (right, bottom)), outline=(0, 0, 255))

        # 使用 UTF-8 编码
        name = name.encode("UTF-8")

        # 在面下方绘制一个带有人名的标签
        text_width, text_height = draw.textsize(name)
        draw.rectangle(((left, bottom - text_height - 10), (right, bottom)),
                       fill=(0, 0, 255), outline=(0, 0, 255))
        draw.text((left + 6, bottom - text_height - 5), name, fill=(255, 255, 255, 255))

    # 从内存中删除图形库
    del draw

    # 显示结果图像
    pil_image.show()


if __name__ == "__main__":
    #步骤 1：训练 KNN 分类器并将其保存到磁盘
    #一旦模型经过训练并保存，下次可以跳过此步骤。
    print("Training KNN classifier...")
    classifier = train("knn_examples/train", model_save_path=
                       "trained_knn_model.clf", n_neighbors=2)
    print("Training complete!")

    #第 2 步：使用训练好的分类器，对未知图像进行预测
    for image_file in os.listdir("knn_examples/test"):
        full_file_path = os.path.join("knn_examples/test", image_file)

        print("Looking for faces in {}".format(image_file))

        # 使用经过训练的分类器模型查找图像中的所有人
        # 可以传入分类器文件名或分类器模型实例
        predictions = predict(full_file_path, model_path="trained_knn_model.clf")

        #在控制台上打印结果
        for name, (top, right, bottom, left) in predictions:
            print("- Found {} at ({}, {})".format(name, left, top))

        #显示覆盖在图像上的结果
        show_prediction_labels_on_image(os.path.join("knn_examples/test",
                                        image_file), predictions)
```

上述代码的算法描述如下。

- KNN 分类器首先训练一组已知的人脸基础上，然后可以预测人脸。
- 在未知图像中寻找 k 个最相似的人脸(欧氏距离下具有相近人脸特征的图像)。
- 在训练集中对标签进行多数投票(可能加权)算法，例如：假如 k=3，那么训练集中与给定图像最接近的 3 张人脸图像就是某人的一张图像。

执行代码后会输出显示如下识别结果。训练 train 目录中的照片，训练完毕后创建训练模型文件 trained_knn_model.clf，然后识别出 test 目录下所有照片的人脸，如图 6-4 所示。

```
Training KNN classifier...
Training complete!
Looking for faces in alex_lacamoire1.jpg
- Found alex_lacamoire at (633, 206)
Looking for faces in johnsnow_test1.jpg
- Found kit_harington at (262, 180)
Looking for faces in kit_with_rose.jpg
- Found rose_leslie at (79, 130)
- Found kit_harington at (247, 92)
Looking for faces in obama1.jpg
- Found obama at (546, 204)
Looking for faces in obama_and_biden.jpg
- Found biden at (737, 449)
- Found obama at (1133, 390)
- Found unknown at (1594, 1062)
```

图 6-4　识别结果

6.3.3 KNN 算法实时识别

请看下面的实例文件 facerec_ipcamera_knn.py，功能是使用 KNN 算法实时识别摄像头中的人脸。本实例的训练算法和预测算法跟前面的实例文件 face_recognition_knn.py 类似，不但可以实时识别摄像头中的人脸，而且可以实时识别网络摄像头中的人脸。

源码路径：daima\6\6-3\facerec_ipcamera_knn.py

```
ALLOWED_EXTENSIONS = {'png', 'jpg', 'jpeg', 'JPG'}

def train(train_dir, model_save_path=None, n_neighbors=None, knn_algo=
        'ball_tree', verbose=False):
    X = []
    y = []
    # 对训练集中的每个人进行循环
    for class_dir in os.listdir(train_dir):
        if not os.path.isdir(os.path.join(train_dir, class_dir)):
            continue

        #循环浏览当前人员的每个训练图像
        for img_path in image_files_in_folder(os.path.join(train_dir, class_dir)):
            image = face_recognition.load_image_file(img_path)
            face_bounding_boxes = face_recognition.face_locations(image)

            if len(face_bounding_boxes) != 1:
                # 如果训练图像中没有人(或人太多)，请跳过该图像。
                if verbose:
                    print("Image {} not suitable for training: {}".format(img_path,
                            "Didn't find a face" if len(face_bounding_boxes) < 1 else
                            "Found more than one face"))
            else:
                # 将当前图像的人脸编码添加到训练集中
                X.append(face_recognition.face_encodings(image, known_face_locations=
                        face_bounding_boxes)[0])
                y.append(class_dir)

    # 确定在 KNN 分类器中用于加权的近邻数
    if n_neighbors is None:
        n_neighbors = int(round(math.sqrt(len(X))))
        if verbose:
            print("Chose n_neighbors automatically:", n_neighbors)
```

```
    # 创建并训练 KNN 分类器
    knn_clf = neighbors.KNeighborsClassifier(n_neighbors=n_neighbors,
                                        algorithm=knn_algo, weights='distance')
    knn_clf.fit(X, y)

    #保存经过训练的 KNN 分类器
    if model_save_path is not None:
        with open(model_save_path, 'wb') as f:
            pickle.dump(knn_clf, f)

    return knn_clf

def predict(X_frame, knn_clf=None, model_path=None, distance_threshold=0.5):
    """
使用训练好的 KNN 分类器识别给定图像中的人脸
    """
    if knn_clf is None and model_path is None:
        raise Exception("Must supply knn classifier either thourgh knn_clf or
                        model_path")

    #加载经过训练的 KNN 模型(如果传入)
    if knn_clf is None:
        with open(model_path, 'rb') as f:
            knn_clf = pickle.load(f)

    X_face_locations = face_recognition.face_locations(X_frame)

    #如果在图像中找不到脸部，则返回空结果
    if len(X_face_locations) == 0:
        return []

    # 在测试图像中查找人脸编码
    faces_encodings = face_recognition.face_encodings(X_frame, known_face_locations=
                                                    X_face_locations)

    #在测试图像中查找人脸编码
    closest_distances = knn_clf.kneighbors(faces_encodings, n_neighbors=1)
    are_matches = [closest_distances[0][i][0] <= distance_threshold for i in
                  range(len(X_face_locations))]

    # 预测类并删除不在阈值内的分类
    return [(pred, loc) if rec else ("unknown", loc) for pred, loc, rec in zip
           (knn_clf.predict(faces_encodings), X_face_locations, are_matches)]
```

```
def show_prediction_labels_on_image(frame, predictions):
    """
    显示人脸识别结果
    """
    pil_image = Image.fromarray(frame)
    draw = ImageDraw.Draw(pil_image)

    for name, (top, right, bottom, left) in predictions:
        # enlarge the predictions for the full sized image.
        top *= 2
        right *= 2
        bottom *= 2
        left *= 2
        #使用 Pillow 模块在脸部周围画一个方框
        draw.rectangle(((left, top), (right, bottom)), outline=(0, 0, 255))
        #使用 UTF-8 编码
        name = name.encode("UTF-8")

        #在面下方绘制一个带有人名的标签
        text_width, text_height = draw.textsize(name)
        draw.rectangle(((left, bottom - text_height - 10), (right, bottom)),
                       fill=(0, 0, 255), outline=(0, 0, 255))
        draw.text((left + 6, bottom - text_height - 5), name, fill=(255, 255, 255, 255))

    # 从内存中删除图形库
    del draw
    #以 open cv 格式保存图像以便能够显示它
    opencvimage = np.array(pil_image)
    return opencvimage

if __name__ == "__main__":
    print("Training KNN classifier...")
    classifier = train("knn_examples/train", model_save_path=
                       "trained_knn_model.clf", n_neighbors=2)
    print("Training complete!")
    #为了提高运行速度，设置每 30 帧处理一帧
    process_this_frame = 29
    print('Setting cameras up...')
    # 如果有多个摄像头，可以用格式 url = 'http://username:password@camera_ip:port'
    #url = 'http://admin:admin@192.168.0.106:8081/'
    cap = cv2.VideoCapture(0)
    while 1 > 0:
        ret, frame = cap.read()
        if ret:
            # 可以根据需要在程序运行时选择不同的调整参数
```

```
# 调整图像大小以获得更稳定的流媒体
img = cv2.resize(frame, (0, 0), fx=0.5, fy=0.5)
process_this_frame = process_this_frame + 1
if process_this_frame % 30 == 0:
    predictions = predict(img, model_path="trained_knn_model.clf")
frame = show_prediction_labels_on_image(frame, predictions)
cv2.imshow('camera', frame)
if ord('q') == cv2.waitKey(10):
    cap.release()
    cv2.destroyAllWindows()
    exit(0)
```

执行代码后可以识别电脑摄像头中的人脸，如图 6-5 所示。

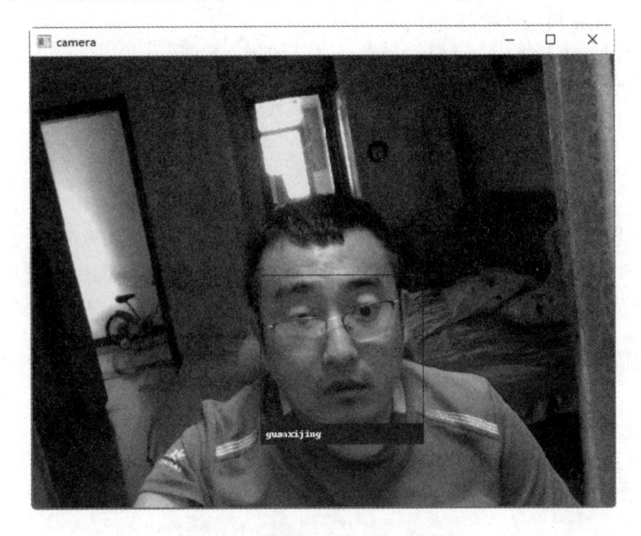

图 6-5　摄像头实时识别结果

第7章

TensorFlow 机器学习和图像识别

TensorFlow 是谷歌公司推出的一个开源库，可以帮助我们开发和训练机器学习模型。在本章的内容中，将详细讲解使用 TensorFlow 机器学习技术实现图像处理的知识，为读者步入本书后面知识的学习打下基础。

7.1 TensorFlow 基础

TensorFlow 是谷歌公司推出的一个开源库，可以帮助我们开发和训练机器学习模型。

扫码观看本节视频讲解

7.1.1 TensorFlow 介绍

TensorFlow 是一个端到端开源机器学习平台。它拥有一个全面而灵活的生态系统，其中包含各种工具、库和社区资源，可助力研究人员推动先进机器学习技术的发展，并使开发者能够轻松地构建和部署由机器学习提供支持的应用。

TensorFlow 由谷歌人工智能团队谷歌大脑(Google Brain)负责开发和维护，拥有包括 TensorFlow Hub、TensorFlow Lite、TensorFlow Research Cloud 在内的多个项目以及各类应用程序接口(Application Programming Interface, API)。自 2015 年 11 月 9 日起，TensorFlow 依据 Apache 2.0 协议开放源代码。

在机器学习框架领域，PyTorch、TensorFlow 已分别成为目前学术界和工业界使用最广泛的两大实力玩家，而紧随其后的 Keras、MXNet 等框架也由于其自身的独特性受到开发者的喜爱。截至 2020 年 8 月，主流机器学习库在 Github 网站活跃度如图 7-1 所示。由此可见，在众多机器学习库中，本章将要讲解的 TensorFlow 最受开发者的欢迎，是当之无愧的机器学习第一库。

	TensorFlow	Keras	MXNet	PyTorch
star	148k	49.4k	18.9k	41.3k
folk	82.5k	18.5k	6.7k	10.8k
contributors	2692	864	828	1540

图 7-1　主流机器学习库的活跃度

7.1.2 TensorFlow 的优势

TensorFlow 是当前最受开发者欢迎的机器学习库，之所以能有现在的地位，主要原因有如下两点。

(1) “背靠大树好乘凉”，Google 几乎在所有应用程序中都使用 TensorFlow 来实现机器学习。得益于 Google 在深度学习领域的影响力和强大的推广能力，TensorFlow 一经推出，关注度就居高不下。

(2) TensorFlow 其本身设计宏大，不仅可以为深度学习提供强有力支持，而且灵活的数值计算核心也能广泛应用于其他涉及大量数学运算的科学领域。

除了上述两点之外，库 TensorFlow 的主要优点如下。

- 支持 Python、JavaScript、C++、Java 和 Go、C＃和 Julia 等多种编程语言。
- 灵活的架构支持多 GPU、分布式训练，跨平台运行能力强。
- 自带 TensorBoard 组件，能够可视化计算图，便于让用户实时监控观察训练过程。
- 官方文档非常详尽，可供开发者查询的资料众多。
- 开发者社区庞大，大量开发者活跃于此，可以共同学习，互相帮助，一起解决学习过程中的问题。

7.1.3 安装 TensorFlow

在安装库 TensorFlow 之前，必须在电脑中安装好 Python。不同版本的 Python，对应需要下载安装的 TensorFlow 的版本也不同。在下载安装 TensorFlow 时，一定要下载正确的版本。例如在笔者电脑中安装的是 Python 3.8，并且操作系统是 64 位的 Windows 10 操作系统，所以我只能安装适应于 Python 3.8 的并且同时适应于 64 位 Windows 系统的 TensorFlow。

1. 使用 pip 安装 TensorFlow

安装 TensorFlow 的最简单方法是使用 pip 命令，在使用这种安装方式时，无须考虑当前所使用的 Python 版本和操作系统的版本，pip 会自动安装适合当前 Python 版本和操作系统版本的 TensorFlow。在安装 Python 后，会自动安装 pip。

(1) 在 Windows 系统中单击左下角的■图标，在弹出列表中找到“命令提示符”，然后用鼠标右键单击“命令提示符”，在弹出的菜单中选择“更多”|“以管理员身份运行”命令，如图 7-2 所示。

(2) 在弹出的“命令提示符”窗口中输入如下命令，即可安装库 TensorFlow：

```
pip install TensorFlow
```

在输入上述 pip 安装命令后，会弹出下载并安装 TensorFlow 的界面，如图 7-3 所示。因为库 TensorFlow 的容量比较大，所以下载过程会比较慢，并且还需要安装相关的其他库，所以整个下载安装过程用时比较长，需要大家耐心等待，确保 TensorFlow 能够正确安装成功。

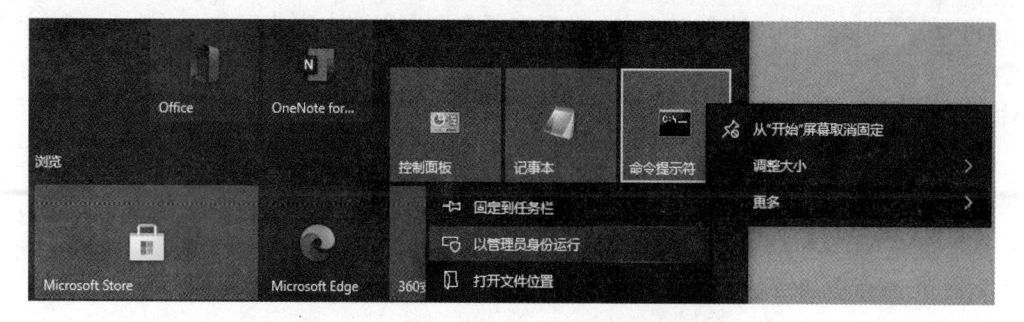

图 7-2 以管理员身份运行“命令提示符”

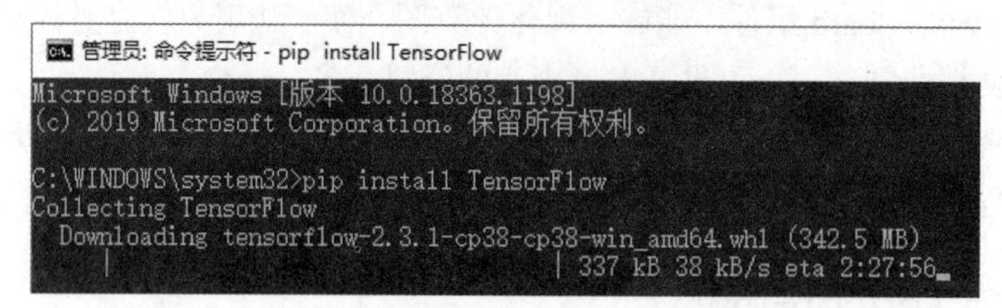

图 7-3 下载、安装 TensorFlow 界面

注意：使用 pip 命令安装的另外一大好处是，自动安装适合当前版本的 TensorFlow。因为在笔者电脑中安装的是 Python 3.8，并且操作系统是 64 位的 Windows 10 操作系统。通过图 7-1 所示的截图可知，这时(我在写作本书时)适合我的最新版本的安装文件是 tensorflow-2.3.7-cp38-cp38-win_amd64.whl。在这个安装文件的名字中，各个字段的含义如下。

- tensorflow-2.3.1：表示 TensorFlow 的版本号是 2.3.1。
- cp38：表示适用于 Python 3.8 版本。
- win_amd64：表示适用于 64 位的 Windows 操作系统。

在使用前面介绍的 pip 方式下载安装 TensorFlow 时，能够安装成功的一个关键因素是网速。如果你的网速过慢，这时候可以考虑在百度中搜索一个 TensorFlow 下载包。因为目前适合我的最新版本的安装文件是 tensorflow-2.3.7-cp38-cp38-win_amd64.whl，那么我可以在百度中搜索这个文件，然后下载。下载完成后保存到本地硬盘中，例如保存位置是 D:\tensorflow-2.3.7-cp38-cp38-win_amd64.whl，那么在“命令提示符”窗口中定位到 D 盘根目录，然后运行如下命令就可以安装 TensorFlow，具体安装过程如图 7-4 所示。

```
pip install tensorflow-2.3.7-cp38-cp38-win_amd64.whl
```

图 7-4　在 Windows 10 的“命令提示符”窗口中安装 TensorFlow

2. 使用 Anaconda 安装 TensorFlow

使用 Anaconda 安装 TensorFlow 的方法和上面介绍的 pip 方式相似，具体流程如下。

(1) 在 Windows 系统中单击左下角的▆图标，在弹出列表中找到 Anaconda Powershell Prompt，然后用鼠标右键单击 Anaconda Powershell Prompt，在弹出的菜单中选择“更多”|“以管理员身份运行”命令，如图 7-5 所示。

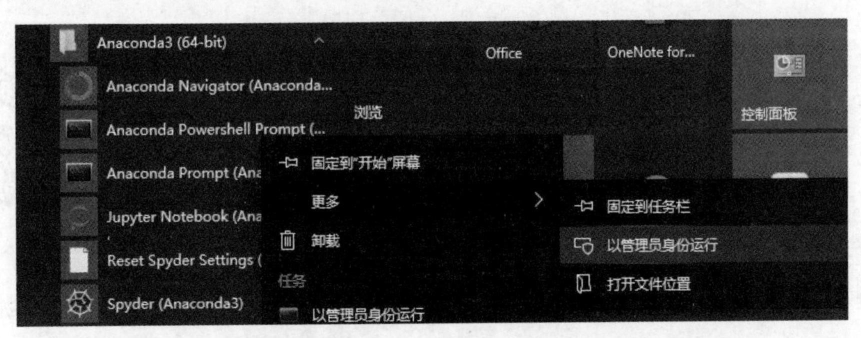

图 7-5　以管理员身份运行“Anaconda Powershell Prompt”

(2) 在弹出的“命令提示符”窗口中输入如下命令，即可安装库 TensorFlow:

```
pip install TensorFlow
```

在输入上述 pip 安装命令后，会弹出下载并安装 TensorFlow 的界面，安装成功后的效果如图 7-6 所示。

```
管理员: Anaconda Powershell Prompt (Anaconda3)
oard<3,>=2.3.0->TensorFlow) (1.7.0)
Requirement already satisfied: requests<3,>=2.21.0 in c:\programdata\anaconda3\lib\site-packages (from tensorboard<3,>=2
.3.0->TensorFlow) (2.22.0)
Requirement already satisfied: markdown>=2.6.8 in c:\programdata\anaconda3\lib\site-packages (from tensorboard<3,>=2.3.0
->TensorFlow) (3.2.2)
Requirement already satisfied: requests-oauthlib>=0.7.0 in c:\programdata\anaconda3\lib\site-packages (from google-auth-
oauthlib<0.5,>=0.4.1->tensorboard<3,>=2.3.0->TensorFlow) (1.3.0)
Requirement already satisfied: cachetools<5.0,>=2.0.0 in c:\programdata\anaconda3\lib\site-packages (from google-auth<2,
>=1.6.3->tensorboard<3,>=2.3.0->TensorFlow) (4.1.1)
Requirement already satisfied: pyasn1-modules>=0.2.1 in c:\programdata\anaconda3\lib\site-packages (from google-auth<2,>
=1.6.3->tensorboard<3,>=2.3.0->TensorFlow) (0.2.8)
Requirement already satisfied: rsa<5,>=3.1.4; python_version >= "3.5" in c:\programdata\anaconda3\lib\site-packages (fro
m google-auth<2,>=1.6.3->tensorboard<3,>=2.3.0->TensorFlow) (4.6)
Requirement already satisfied: urllib3!=1.25.0,!=1.25.1,<1.26,>=1.21.1 in c:\programdata\anaconda3\lib\site-packages (fr
om requests<3,>=2.21.0->tensorboard<3,>=2.3.0->TensorFlow) (1.25.8)
Requirement already satisfied: certifi>=2017.4.17 in c:\programdata\anaconda3\lib\site-packages (from requests<3,>=2.21.
0->tensorboard<3,>=2.3.0->TensorFlow) (2019.11.28)
Requirement already satisfied: chardet<3.1.0,>=3.0.2 in c:\programdata\anaconda3\lib\site-packages (from requests<3,>=2.
21.0->tensorboard<3,>=2.3.0->TensorFlow) (3.0.4)
Requirement already satisfied: idna<2.9,>=2.5 in c:\programdata\anaconda3\lib\site-packages (from requests<3,>=2.21.0->t
ensorboard<3,>=2.3.0->TensorFlow) (2.8)
Requirement already satisfied: importlib-metadata; python_version < "3.8" in c:\programdata\anaconda3\lib\site-packages
(from markdown>=2.6.8->tensorboard<3,>=2.3.0->TensorFlow) (1.5.0)
Requirement already satisfied: oauthlib>=3.0.0 in c:\programdata\anaconda3\lib\site-packages (from requests-oauthlib>=0.
7.0->google-auth-oauthlib<0.5,>=0.4.1->tensorboard<3,>=2.3.0->TensorFlow) (3.1.0)
Requirement already satisfied: pyasn1<0.5.0,>=0.4.6 in c:\programdata\anaconda3\lib\site-packages (from pyasn1-modules>=
0.2.1->google-auth<2,>=1.6.3->tensorboard<3,>=2.3.0->TensorFlow) (0.4.8)
Requirement already satisfied: zipp>=0.5 in c:\programdata\anaconda3\lib\site-packages (from importlib-metadata; python_
version < "3.8"->markdown>=2.6.8->tensorboard<3,>=2.3.0->TensorFlow) (2.2.0)
(base) PS C:\WINDOWS\system32>
```

图 7-6　下载、安装 TensorFlow 界面

7.2　创建第一个机器学习程序

在本节的内容中，将使用 TensorFlow 编写第一个机器学习程序，并分别在 PyCharm 环境和 Colaboratory 环境调试运行这个程序。

扫码观看本节视频讲解

7.2.1　在 PyCharm 环境实现

首先打开 PyCharm，然后新建一个名为 TF01 的 Python 工程。然后在工程 TF01 中新建一个 Python 程序文件 tf01.py，实例文件 tf01.py 的具体实现流程如下。

源码路径：daima\7\7-2\TF01\tf01.py

(1)　使用 import 语句导入库 tensorflow，并将库 tensorflow 简写为 tf。代码如下：

```
import tensorflow as tf
```

(2) 加载 MNIST 数据集，并将样本从整数转换为浮点数。代码如下：

```
mnist = tf.keras.datasets.mnist
 (x_train, y_train), (x_test, y_test) = mnist.load_data()
```

(3) 对数据进行归一化处理。像素范围是 0～255，所以都除以 255，归一化到(0，1)之间。代码如下：

```
x_train, x_test = x_train / 255.0, x_test / 255.0
```

在库 TensorFlow 中内置了 MNIST 数据集，所以可以直接使用上述代码导入 MNIST 数据集。

(4) 将 model 模型的各层堆叠起来，以搭建 tf.keras.Sequential 模型。代码如下：

```
model = tf.keras.models.Sequential([
  tf.keras.layers.Flatten(input_shape=(28, 28)),
  tf.keras.layers.Dense(128, activation='relu'),
  tf.keras.layers.Dropout(0.2),
  tf.keras.layers.Dense(10, activation='softmax')
])
```

代码说明如下。

- tf.keras.layers.Flatten(input_shape=(28, 28))：用于添加 Flatten 层，将数据变成[28×28]格式。
- tf.keras.layers.Dense(128, activation='relu')：函数 Dense()是定义网络层的基本方法，此行代码的功能是将网络层设置为 128 个。
- tf.keras.layers.Dropout(0.2)：在深度学习网络的训练过程中，对于神经网络单元，按照一定的概率将其暂时从网络中丢弃，这样可以防止过拟合。上述功能是通过函数 Dropout()实现的，此行代码的功能是按照 0.2 的概率丢弃神经网络单元。
- tf.keras.layers.Dense(10, activation='softmax')：使用函数 Dense()修改网络层的个数。

(5) 为机器学习训练选择优化器和损失函数，代码如下：

```
model.compile(optimizer='adam',
             loss='sparse_categorical_crossentropy',
             metrics=['accuracy'])
```

函数 model.compile()用于设置训练方法的参数，以及在训练时用的优化器、损失函数和准确率评测标准。

- optimizer：用于设置优化器，可以是字符串形式的优化器名字，也可以是函数形式。使用函数形式可以设置学习率、动量和超参数。

- loss：用于设置损失函数，可以是字符串形式的损失函数的名字，也可以是函数形式。
- metrics：表示准确率。

(6) 训练并验证模型，代码如下：

```
model.fit(x_train, y_train, epochs=5)
model.evaluate(x_test,  y_test, verbose=2)
```

代码说明如下。

- model.fit(x_train, y_train, epochs=5)：功能是根据样本进行训练，通过函数 fit()展示训练过程，包括损失函数和其他指标的数值随 epochs(训练总轮数)值而变化的情况。
- model.evaluate(x_test, y_test, verbose=2)：使用函数 evaluate()评估已经训练过的模型，分别返回损失值(loss)和准确率(accuracy)。

运行实例文件 tf01.py，执行完毕后会得到下面的结果：

```
Epoch 1/5
1875/1875 [==============================] - 5s 2ms/step - loss: 0.3014 -
accuracy: 0.9120
Epoch 2/5
1875/1875 [==============================] - 5s 3ms/step - loss: 0.1455 -
accuracy: 0.9567
Epoch 3/5
1875/1875 [==============================] - 5s 3ms/step - loss: 0.1065 -
accuracy: 0.9672
Epoch 4/5
1875/1875 [==============================] - 5s 3ms/step - loss: 0.0893 -
accuracy: 0.9723
Epoch 5/5
1875/1875 [==============================] - 5s 3ms/step - loss: 0.0761 -
accuracy: 0.9761
313/313 - 0s - loss: 0.0762 - accuracy: 0.9762
```

通过上述执行结果可知，我们训练模型的精确度高达 97.62%，这是一个非常优秀的结果。

7.2.2 在 Colaboratory 环境实现

(1) 通过 Google Chrome 浏览器登录 Colaboratory 云端服务器，选择“文件”|“新建笔记本”命令，创建一个新的 Jupyter Notebook 文件，将文件命名为 tf02.ipynb。

(2) 在文件 tf02.ipynb 中编写如下所示的代码。

源码路径：daima\7\7-2\TF01\tf02.ipynb

```
import tensorflow as tf

mnist = tf.keras.datasets.mnist
(x_train, y_train), (x_test, y_test) = mnist.load_data()
x_train, x_test = x_train / 255.0, x_test / 255.0

model = tf.keras.models.Sequential([
  tf.keras.layers.Flatten(input_shape=(28, 28)),
  tf.keras.layers.Dense(128, activation='relu'),
  tf.keras.layers.Dropout(0.2),
  tf.keras.layers.Dense(10, activation='softmax')
])

model.compile(optimizer='adam',
              loss='sparse_categorical_crossentropy',
              metrics=['accuracy'])

model.fit(x_train, y_train, epochs=5)
model.evaluate(x_test,  y_test, verbose=2)
```

(3) 使用 Colaboratory 加速器运行实例文件 tf02.ipynb，选择“代码执行程序”|“更改运行时类型”命令，如图 7-7 所示。在弹出的对话框中可以选择硬件加速器，例如选择 GPU，然后单击“保存”按钮，如图 7-8 所示。

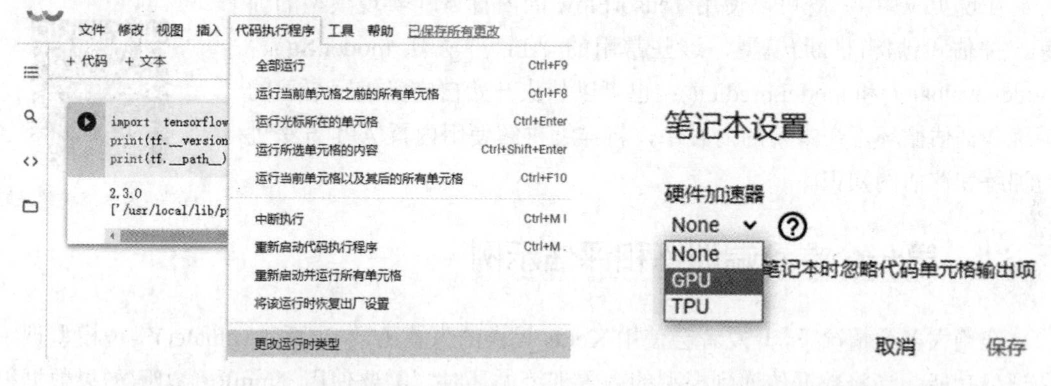

图 7-7 选择“更改运行时类型” 图 7-8 选择加速类型

(4) 单击按钮运行文件 tf02.ipynb，执行效果如图 7-9 所示。

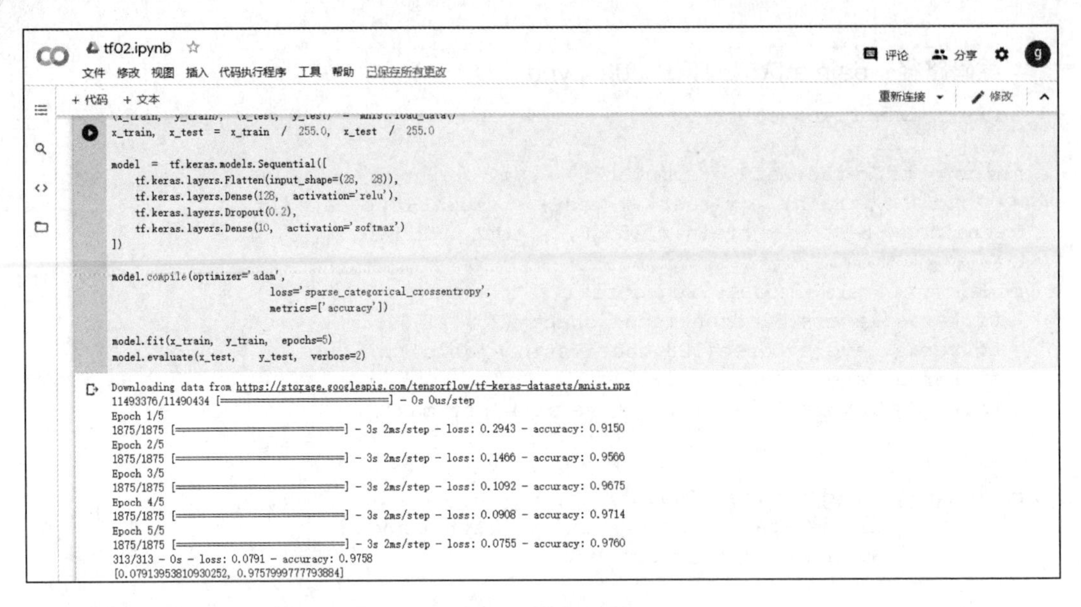

图 7-9 执行效果

7.3 使用内置方法进行训练和评估

在现实应用中，既可以使用 TensorFlow 的内置 API 实现模型的训练、评估和预测(推断)模型，这些常用的 API 方法有 model.fit()、model.evaluate()和 model.predict()；也可以从头开始自定义编写自己的训练和评估循环。在本节的内容中，将详细讲解使用内置 API 方法进行训练和评估的知识。

扫码观看本节视频讲解

7.3.1 第一个端到端训练和评估示例

在绝大多数情况下，开发者会使用 Keras 提供的内置函数 fit()和 evaluate()实现模型训练和评估功能。当将数据传递到模型的内置训练循环时，需要使用 NumPy 数组(如果数据很小且适合装入内存)或 tf.data Dataset 对象进行处理。

在下面的实例中，将 MNIST 数据集作为 NumPy 数组，以演示使用优化器、损失和指标进行训练和评估的方法。

实例文件 xun01.py 的具体实现流程如下所示。

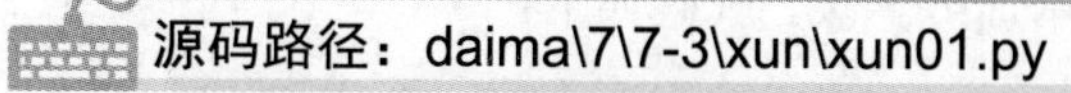

(1) 使用 import 语句导入需要的 keras 模块，代码如下：

```
from tensorflow import keras
from tensorflow.keras import layers
```

(2) 使用函数式 API 构建模型，代码如下：

```
inputs = keras.Input(shape=(784,), name="digits")
x = layers.Dense(64, activation="relu", name="dense_1")(inputs)
x = layers.Dense(64, activation="relu", name="dense_2")(x)
outputs = layers.Dense(10, activation="softmax", name="predictions")(x)

model = keras.Model(inputs=inputs, outputs=outputs)
```

(3) 加载 MNIST 数据集中的图像数据，将这些数据作为测试数据进行评估。代码如下：

```
(x_train, y_train), (x_test, y_test) = keras.datasets.mnist.load_data()

# 预处理数据(这些都是 NumPy 数组)
x_train = x_train.reshape(60000, 784).astype("float32") / 255
x_test = x_test.reshape(10000, 784).astype("float32") / 255

y_train = y_train.astype("float32")
y_test = y_test.astype("float32")

#保留 10000 个样品进行验证
x_val = x_train[-10000:]
y_val = y_train[-10000:]
x_train = x_train[:-10000]
y_train = y_train[:-10000]
```

(4) 使用函数 compile()编译上面的模型，在 compile()中设置训练配置信息，包括优化器、损失和指标等参数。代码如下：

```
model.compile(
    optimizer=keras.optimizers.RMSprop(),  # 优化器
    # 最小化损失函数
    loss=keras.losses.SparseCategoricalCrossentropy(),
    # 要监视的指标列表
    metrics=[keras.metrics.SparseCategoricalAccuracy()],
)
```

(5) 使用函数 fit()在数据上拟合模型。函数 fit()会将数据切成大小为 batch_size 的“批次”，然后将整个数据集重复迭代指定数量的“周期”来训练模型。其中 epochs 表示迭代

次数，1 个 epoch 等于使用训练集中的全部样本训练一次。代码如下：

```
print("Fit model on training data")
history = model.fit(
    x_train,
    y_train,
    batch_size=64,
    epochs=2,
    # 通过一些验证来监控每个 epoch 结束时的验证损失和度量
    validation_data=(x_val, y_val),
)
```

(6) 使用 print()打印输出 history 对象的值，history 对象的功能是保留训练期间的损失值和指标值记录。代码如下：

```
print(history.history)
```

(7) 使用 evaluate()基于测试数据来评估模型，代码如下：

```
print("数据测试评估")
results = model.evaluate(x_test, y_test, batch_size=128)
print("Loss, Acc", results)
```

(8) 使用 predict()基于新数据生成预测，代码如下：

```
print("生成 3 个样本的预测")
predictions = model.predict(x_test[:3])
print("预测:", predictions.shape)
```

在作者电脑中执行代码后会输出：

```
Fit model on training data
Epoch 1/2
782/782 [==============================] - 2s 3ms/step - loss: 0.3416 -
sparse_categorical_accuracy: 0.9032 - val_loss: 0.1813 -
val_sparse_categorical_accuracy: 0.9482
Epoch 2/2
782/782 [==============================] - 2s 2ms/step - loss: 0.1619 -
sparse_categorical_accuracy: 0.9518 - val_loss: 0.1363 -
val_sparse_categorical_accuracy: 0.9587
{'loss': [0.3415972590446472, 0.16185878217220306],
'sparse_categorical_accuracy': [0.9031599760055542, 0.9518200159072876],
'val_loss': [0.18128132820129395, 0.1362582892179489],
'val_sparse_categorical_accuracy': [0.948199987411499, 0.9587000012397766]}
数据测试评估
79/79 [==============================] - 0s 2ms/step - loss: 0.1398 -
sparse_categorical_accuracy: 0.9574
Loss, Acc [0.13983243703842163, 0.9574000239372253]
```

```
生成 3 个样本的预测
预测: (3, 10)
```

7.3.2 使用 compile()训练模型

在 TensorFlow 应用中，要想使用方法 compile()训练模型，需要设置损失、指标、优化器以及一些要监视的指标(可选)等信息。这些信息被作为方法 compile()的参数传递给模型，例如：

```
model.compile(
    optimizer=keras.optimizers.RMSprop(learning_rate=1e-3),
    loss=keras.losses.SparseCategoricalCrossentropy(),
    metrics=[keras.metrics.SparseCategoricalAccuracy()],
)
```

参数说明如下。

(1) 参数 optimizer：这是一个优化参数，用于设置选择范围，可选的值中包括下面的成员：['Adadelta', 'Adagrad', 'Adam', 'Adamax', 'FTRL', 'NAdam', 'optimizer', 'RMSprop', 'SGD']。

在使用参数 optimizer 时，可以使用字符串形式给出的优化器名字，也可以使用函数形式。在使用函数形式时，可以设置学习率、动量和超参数，例如在如下代码中通过字符串方式使用了 SGD 优化器：

```
model.compile(
    optimizer=" SGD ",
)
```

而在如下代码中通过函数形式使用了 SGD 优化器：

```
model.compile(
    optimizer= tf.optimizers.SGD(lr = 学习率, decay = 学习率衰减率, momentum = 动量参数),
)
```

(2) 参数 metrics：表示在训练和测试期间要使用的评估指标，如果模型具有多个输出，则可以为每个输出设置不同的损失和指标，并且可以调节每个输出对模型总损失的贡献。

(3) 参数 loss：表示损失值，可以是字符串形式给出的损失函数的名字，也可以是函数形式。可以为参数 loss 设置如下字符串形式的损失函数名。

- BinaryCrossentropy：计算真实标签和预测标签之间的交叉熵损失。
- CategoricalCrossentropy：计算标签和预测之间的交叉熵损失，要求目标为 onehot 编码。

- CategoricalHinge：计算 y_true 和 y_pred 之间的分类铰链损失。
- CosineSimilarity：计算 y_true 和 y_pred 之间的余弦相似度。
- Hinge：计算 y_true 和 y_pred 之间的铰链损耗。
- Huber：计算 y_true 和 y_pred 之间的 Huber 损失。
- KLDivergence：计算 y_true 和 y_pred 之间的 Kullback Leibler 差异损失。
- LogCosh：计算预测误差的双曲余弦的对数。
- Loss：损失基类。
- MeanAbsoluteError：计算标签和预测之间的绝对差异的平均值。
- MeanAbsolutePercentageError：计算 y_true 和 y_pred 之间的平均绝对百分比误差。
- MeanSquaredError：计算标签和预测之间的误差平方的平均值。
- MeanSquaredLogarithmicError：计算 y_true 和 y_pred 之间的均方对数误差。
- Poisson：计算 y_true 和 y_pred 之间的泊松损失。
- Reduction：减少损失的类型。
- SparseCategoricalCrossentropy：计算标签和预测之间的交叉熵损失，要求目标为非 onehot 编码，在函数内部实现 onehot 编码。
- SquaredHinge：计算 y_true 和 y_pred 之间的平方铰链损耗。

如果开发者倾向于使用上述参数的默认设置，那么可以通过字符串标识符的方式快速设置优化器、损失和评估指标，例如下面这种简洁的设置方式：

```
model.compile(
    optimizer="rmsprop",
    loss="sparse_categorical_crossentropy",
    metrics=["sparse_categorical_accuracy"],
)
```

为了方便在以后重用训练配置信息，可以将模型的定义和编译步骤放入到函数中。例如下面的代码：

```
def get_uncompiled_model():
    inputs = keras.Input(shape=(784,), name="digits")
    x = layers.Dense(64, activation="relu", name="dense_1")(inputs)
    x = layers.Dense(64, activation="relu", name="dense_2")(x)
    outputs = layers.Dense(10, activation="softmax", name="predictions")(x)
    model = keras.Model(inputs=inputs, outputs=outputs)
    return model

def get_compiled_model():
    model = get_uncompiled_model()
    model.compile(
```

```
        optimizer="rmsprop",
        loss="sparse_categorical_crossentropy",
        metrics=["sparse_categorical_accuracy"],
    )
    return model
```

7.3.3 自定义损失

在现实应用中，有两种使用 Keras 自定义损失的方法。

1) 第一种方式

例如在下面的实例中创建了函数 custom_mean_squared_error()，此函数的功能是计算实际数据与预测值之间的均方误差的损失，此函数接收的参数是 y_true 和 y_pred。

实例文件 xun02.py 的具体实现流程如下所示。

源码路径：daima\7\7-3\xun\xun02.py

(1) 加载 MNIST 数据集中的图像数据，将其作为测试数据进行评估。代码如下：

```
(x_train, y_train), (x_test, y_test) = keras.datasets.mnist.load_data()

# 预处理数据(这些都是 Numpy 数组)
x_train = x_train.reshape(60000, 784).astype("float32") / 255
x_test = x_test.reshape(10000, 784).astype("float32") / 255

y_train = y_train.astype("float32")
y_test = y_test.astype("float32")

#保留 10000 个样品进行验证
x_val = x_train[-10000:]
y_val = y_train[-10000:]
x_train = x_train[:-10000]
y_train = y_train[:-10000]
```

(2) 为了便于在以后的项目中方便调用我们的模型，分别定义专用函数实现模型定义和编译功能。其中在函数 get_uncompiled_model() 中实现模型定义，在函数 get_compiled_model()中实现模型编译。代码如下：

```
def get_uncompiled_model():
    inputs = keras.Input(shape=(784,), name="digits")
    x = layers.Dense(64, activation="relu", name="dense_1")(inputs)
    x = layers.Dense(64, activation="relu", name="dense_2")(x)
    outputs = layers.Dense(10, activation="softmax", name="predictions")(x)
    model = keras.Model(inputs=inputs, outputs=outputs)
```

```
    return model


def get_compiled_model():
    model = get_uncompiled_model()
    model.compile(
        optimizer="rmsprop",
        loss="sparse_categorical_crossentropy",
        metrics=["sparse_categorical_accuracy"],
    )
    return model
```

(3) 使用函数 custom_mean_squared_error()计算损失，代码如下：

```
def custom_mean_squared_error(y_true, y_pred):
    return tf.math.reduce_mean(tf.square(y_true - y_pred))
```

(4) 调用模型进行训练，并打印输出计算的损失值，代码如下：

```
model = get_uncompiled_model()
model.compile(optimizer=keras.optimizers.Adam(),
loss=custom_mean_squared_error)

y_train_one_hot = tf.one_hot(y_train, depth=10)
model.fit(x_train, y_train_one_hot, batch_size=64, epochs=1)
```

在作者电脑中执行代码后会输出：

```
782/782 [==============================] - 1s 2ms/step - loss: 0.0159
<tensorflow.python.keras.callbacks.History at 0x7f01501a66a0>
```

2) 第 2 种方式

如果需要实现一个使用除 y_true 和 y_pred 之外的其他参数的损失函数，则可以将类 tf.keras.losses.Loss 进行子类化处理，并实现如下两个方法。

- __init__(self)：接收在调用损失函数期间传递的参数。
- call(self, y_true, y_pred)：使用目标(y_true)和模型预测(y_pred)来计算模型的损失。

假设想要使用均方误差，但存在一个会抑制预测值偏离 0.5(假设分类目标采用独热编码，且取值介于 0 和 1 之间)的附加项。我们可以为模型创建一个激励，使其不会对预测值过于自信，这可能对减轻过拟合有帮助。

实例文件 xun03.py 的主要实现代码如下所示，演示了将类 tf.keras.losses.Loss 进行子类化处理的用法。

源码路径：daima\7\7-3\xun\xun03.py

```
class CustomMSE(keras.losses.Loss):
    def __init__(self, regularization_factor=0.1, name="custom_mse"):
        super().__init__(name=name)
        self.regularization_factor = regularization_factor

    def call(self, y_true, y_pred):
        mse = tf.math.reduce_mean(tf.square(y_true - y_pred))
        reg = tf.math.reduce_mean(tf.square(0.5 - y_pred))
        return mse + reg * self.regularization_factor

def get_uncompiled_model():
    inputs = keras.Input(shape=(784,), name="digits")
    x = layers.Dense(64, activation="relu", name="dense_1")(inputs)
    x = layers.Dense(64, activation="relu", name="dense_2")(x)
    outputs = layers.Dense(10, activation="softmax", name="predictions")(x)
    model = keras.Model(inputs=inputs, outputs=outputs)
    return model

def get_compiled_model():
    model = get_uncompiled_model()
    model.compile(
        optimizer="rmsprop",
        loss="sparse_categorical_crossentropy",
        metrics=["sparse_categorical_accuracy"],
    )
    return model

model = get_uncompiled_model()
model.compile(optimizer=keras.optimizers.Adam(), loss=CustomMSE())

y_train_one_hot = tf.one_hot(y_train, depth=10)
model.fit(x_train, y_train_one_hot, batch_size=64, epochs=1)
```

在作者电脑中执行代码后会输出：

```
782/782 [==============================] - 1s 2ms/step - loss: 0.0387
<tensorflow.python.keras.callbacks.History at 0x7f01500b4a90>
```

7.3.4 自定义指标

如果不想使用 API 提供的内置指标，也可以通过将类 tf.keras.metrics.Metric 子类化的方式来创建自定义指标，此时需要实现如下 4 个方法。

- __init__(self)：将在其中为指标创建状态变量。
- update_state(self, y_true, y_pred, sample_weight=None)：使用目标 y_true 和模型预测 y_pred 更新状态变量。
- result(self)：使用状态变量来计算最终结果。
- reset_states(self)：用于重新初始化指标的状态。

其中 update_state()和 result()被分别保存在 update_state()和 result()中，这是因为在某些情况下，result()的开销可能会非常大，并且只能定期执行。

例如下面是一个实现 CategoricalTruePositives 指标的例子，此指标可以计算有多少样本被正确分类为指定的类。

实例文件 xun04.py 的主要实现代码如下所示。

源码路径：daima\7\7-3\xun\xun04.py

```
class CategoricalTruePositives(keras.metrics.Metric):
    def __init__(self, name="categorical_true_positives", **kwargs):
        super(CategoricalTruePositives, self).__init__(name=name, **kwargs)
        self.true_positives = self.add_weight(name="ctp", initializer="zeros")

    def update_state(self, y_true, y_pred, sample_weight=None):
        y_pred = tf.reshape(tf.argmax(y_pred, axis=1), shape=(-1, 1))
        values = tf.cast(y_true, "int32") == tf.cast(y_pred, "int32")
        values = tf.cast(values, "float32")
        if sample_weight is not None:
            sample_weight = tf.cast(sample_weight, "float32")
            values = tf.multiply(values, sample_weight)
        self.true_positives.assign_add(tf.reduce_sum(values))

    def result(self):
        return self.true_positives

    def reset_states(self):
        # The state of the metric will be reset at the start of each epoch
        self.true_positives.assign(0.0)


model = get_uncompiled_model()
model.compile(
    optimizer=keras.optimizers.RMSprop(learning_rate=1e-3),
    loss=keras.losses.SparseCategoricalCrossentropy(),
    metrics=[CategoricalTruePositives()],
)
model.fit(x_train, y_train, batch_size=64, epochs=3)
```

在作者电脑中执行代码后会输出：

```
Epoch 1/3
782/782 [==============================] - 5s 6ms/step - loss: 0.3436 -
categorical_true_positives: 45098.0000
Epoch 2/3
782/782 [==============================] - 4s 5ms/step - loss: 0.1661 -
categorical_true_positives: 47544.0000
Epoch 3/3
782/782 [==============================] - 4s 5ms/step - loss: 0.1223 -
categorical_true_positives: 48164.0000
<tensorflow.python.keras.callbacks.History at 0x7f0750069e10>
```

7.3.5 处理不适合标准签名的损失和指标

虽然我们可以根据 y_true 和 y_pred 计算出绝大多数的损失和指标(其中 y_pred 表示模型的输出)，但是仍然有一些损失和指标无法计算。例如在现实应用中，正则化损失可能需要用到激活层(这种情况下没有目标)，并且此激活可能不是模型输出。在这种情况下，可以在自定义层的内部调用方法 self.add_loss(loss_value)，以这种方式添加的损失会在训练期间添加到“主要”损失中(传递给 compile() 的损失)。下面是一个添加激活正则化的例子，激活正则化被内置于所有的 Keras 层中。

实例文件 xun05.py 的主要实现代码如下所示。

源码路径：daima\7\7-3\xun\xun05.py

```
class ActivityRegularizationLayer(layers.Layer):
   def call(self, inputs):
      self.add_loss(tf.reduce_sum(inputs) * 0.1)
      return inputs  #通过层

inputs = keras.Input(shape=(784,), name="digits")
x = layers.Dense(64, activation="relu", name="dense_1")(inputs)

# 将活动正则化作为层来插入
x = ActivityRegularizationLayer()(x)

x = layers.Dense(64, activation="relu", name="dense_2")(x)
outputs = layers.Dense(10, name="predictions")(x)

model = keras.Model(inputs=inputs, outputs=outputs)
model.compile(
   optimizer=keras.optimizers.RMSprop(learning_rate=1e-3),
```

```
    loss=keras.losses.SparseCategoricalCrossentropy(from_logits=True),
)

# 由于正则化分量的存在，显示的损耗将会比以前大得多。
model.fit(x_train, y_train, batch_size=64, epochs=1)
```

在作者电脑中执行代码后会输出：

```
782/782 [==============================] - 1s 2ms/step - loss: 2.4796
<tensorflow.python.keras.callbacks.History at 0x7f073c2229b0>
```

另外，也可以使用函数add_metric()对记录指标值执行相同的操作：

```
class MetricLoggingLayer(layers.Layer):
    def call(self, inputs):
        #参数aggregation用于设置如何在每个epoch上聚合per-batch值
        #在这种情况下，我们简单地求平均值
        self.add_metric(
            keras.backend.std(inputs), name="std_of_activation", aggregation="mean"
        )
        return inputs # Pass-through layer.


inputs = keras.Input(shape=(784,), name="digits")
x = layers.Dense(64, activation="relu", name="dense_1")(inputs)

#插入std日志作为一个层。
x = MetricLoggingLayer()(x)

x = layers.Dense(64, activation="relu", name="dense_2")(x)
outputs = layers.Dense(10, name="predictions")(x)

model = keras.Model(inputs=inputs, outputs=outputs)
model.compile(
    optimizer=keras.optimizers.RMSprop(learning_rate=1e-3),
    loss=keras.losses.SparseCategoricalCrossentropy(from_logits=True),
)
model.fit(x_train, y_train, batch_size=64, epochs=1)
```

此时执行代码后会输出：

```
782/782 [==============================] - 2s 2ms/step - loss: 0.3423 -
std_of_activation: 1.0018
<tensorflow.python.keras.callbacks.History at 0x7f073c07ae80>
```

在函数式API中，还可以调用函数model.add_loss(loss_tensor)或model.add_metric(metric_tensor, name, aggregation)实现类似的功能，例如下面的代码：

```
inputs = keras.Input(shape=(784,), name="digits")
x1 = layers.Dense(64, activation="relu", name="dense_1")(inputs)
x2 = layers.Dense(64, activation="relu", name="dense_2")(x1)
outputs = layers.Dense(10, name="predictions")(x2)
model = keras.Model(inputs=inputs, outputs=outputs)

model.add_loss(tf.reduce_sum(x1) * 0.1)

model.add_metric(keras.backend.std(x1), name="std_of_activation",
                 aggregation="mean")

model.compile(
   optimizer=keras.optimizers.RMSprop(1e-3),
   loss=keras.losses.SparseCategoricalCrossentropy(from_logits=True),
)
model.fit(x_train, y_train, batch_size=64, epochs=1)
```

此时执行代码后会输出：

```
782/782 [==============================] - 2s 2ms/step - loss: 2.4963 -
std_of_activation: 0.0019
<tensorflow.python.keras.callbacks.History at 0x7f06fc62c630>
```

当通过函数 add_loss()传递损失时，可以在没有损失函数的情况下调用 compile()，因为模型已经有损失要最小化。

实例文件 xun06.py 的具体实现流程如下所示。

源码路径：daima\7\7-3\xun\xun06.py

(1) 新建层 LogisticEndpoint，以目标和 logits 作为输入，并通过函数 add_loss()跟踪交叉熵损失。另外，LogisticEndpoint 还通过函数 add_metric()跟踪分类准确率。代码如下：

```
class LogisticEndpoint(keras.layers.Layer):
   def __init__(self, name=None):
      super(LogisticEndpoint, self).__init__(name=name)
      self.loss_fn = keras.losses.BinaryCrossentropy(from_logits=True)
      self.accuracy_fn = keras.metrics.BinaryAccuracy()

   def call(self, targets, logits, sample_weights=None):
      #计算训练时间损失值，并使用 self.add_loss()将值添加到层
      loss = self.loss_fn(targets, logits, sample_weights)
      self.add_loss(loss)

      # 将精度记录为度量，并使用 self.add_metric()将其添加到层中
      acc = self.accuracy_fn(targets, logits, sample_weights)
      self.add_metric(acc, name="accuracy")
```

```
    # 返回推理时间预测张量 (为'.predict()').
    return tf.nn.softmax(logits)
```

(2) 在具有两个输入(输入数据和目标)的模型中使用层 LogisticEndpoint，在编译时无须使用 loss 参数。代码如下：

```
import numpy as np

inputs = keras.Input(shape=(3,), name="inputs")
targets = keras.Input(shape=(10,), name="targets")
logits = keras.layers.Dense(10)(inputs)
predictions = LogisticEndpoint(name="predictions")(logits, targets)

model = keras.Model(inputs=[inputs, targets], outputs=predictions)
model.compile(optimizer="adam")  #没有 loss 参数

data = {
    "inputs": np.random.random((3, 3)),
    "targets": np.random.random((3, 10)),
}
model.fit(data)
```

在作者电脑中执行代码后会输出：

```
1/1 [==============================] - 0s 2ms/step - loss: 0.9475 -
binary_accuracy: 0.0000e+00
<tensorflow.python.keras.callbacks.History at 0x7f075c10eeb8>
```

7.3.6 自动分离验证预留集

在本章前面第一个端到端实例文件 tf01.py 中，在使用函数 tf.keras.Model.fit()时，用参数 validation_data 将 NumPy 数组(x_val, y_val)的元组传递给模型，用于在每个周期结束时评估验证损失和验证指标。除此之外，函数 tf.keras.Model.fit()的另一个参数 validation_split 允许自动保留部分训练数据以供验证。也就是说，分割部分数据用于验证，其余用于训练。参数 validation_split 的值表示要保留用于验证的数据比例，因此应将其设置为大于 0 且小于 1 的数字。例如：

- validation_split=0.2 表示“使用 20% 的数据进行验证”。
- validation_split=0.6 表示“使用 60% 的数据进行验证”。

验证数据的计算方式为：抽取通过 tf.keras.Model.fit()调用接收的数组的最后 x%的样本。请注意，仅在使用 NumPy 数据进行训练时才能使用参数 validation_split。也就是说，当 x 是 dataset、dataset iterator、generator 或 keras.utils.Sequence 时该参数不可用。

实例文件 xun07.py 的主要实现代码如下所示。

源码路径：daima\7\7-3\xun\xun07.py

```
def get_uncompiled_model():
    inputs = keras.Input(shape=(784,), name="digits")
    x = layers.Dense(64, activation="relu", name="dense_1")(inputs)
    x = layers.Dense(64, activation="relu", name="dense_2")(x)
    outputs = layers.Dense(10, activation="softmax", name="predictions")(x)
    model = keras.Model(inputs=inputs, outputs=outputs)
    return model

def get_compiled_model():
    model = get_uncompiled_model()
    model.compile(
        optimizer="rmsprop",
        loss="sparse_categorical_crossentropy",
        metrics=["sparse_categorical_accuracy"],
    )
    return model

model = get_compiled_model()
model.fit(x_train, y_train, batch_size=64, validation_split=0.2, epochs=1)
```

在作者电脑中执行代码后会输出：

```
11493376/11490434 [==========================] - 0s 0us/step
625/625 [==========================] - 2s 3ms/step - loss: 0.3712 - sparse_categorical_
accuracy: 0.8936 - val_loss: 0.2212 - val_sparse_categorical_accuracy: 0.9327
<tensorflow.python.keras.callbacks.History at 0x7f69921a5fd0>
```

7.3.7 通过 tf.data 数据集进行训练和评估

经过本章前面内容的学习，大家已经基本上了解了处理损失、指标和优化器的知识，并且已经掌握了将数据作为 NumPy 数组传递时，在函数 tf.keras.Model.fit()中使用参数 validation_data 和参数 validation_split 设置配置信息的方法。

从 TensorFlow 2.0 开始，提供了专门用于实现数据输入的接口 tf.data.Dataset，它能够以快速且可扩展的方式加载和预处理数据，帮助开发者高效地实现数据的读入、打乱(shuffle)、增强(augment)等功能。

实例文件 xun08.py 的主要实现代码如下所示，演示了使用 tf.data 数据集进行训练和评估的过程。

源码路径：daima\7\7-3\xun\xun08.py

```
# 首先创建一个训练数据集实例，接下来将使用与前面相同的 MNIST 数据
train_dataset = tf.data.Dataset.from_tensor_slices((x_train, y_train))
# 洗牌并切片数据集.
train_dataset = train_dataset.shuffle(buffer_size=1024).batch(64)

# 现在得到了一个测试数据集
test_dataset = tf.data.Dataset.from_tensor_slices((x_test, y_test))
test_dataset = test_dataset.batch(64)

#由于数据集已经处理批处理，所以我们不传递“batch\u size”参数
model.fit(train_dataset, epochs=3)

#还可以对数据集进行评估或预测
print("Evaluate 评估:")
result = model.evaluate(test_dataset)
dict(zip(model.metrics_names, result))
```

在上述代码中，使用 dataset 的内置函数 shuffle()将数据打乱，此函数的参数值越大，混乱程度就越大。另外，还可以使用 dataset 的其他内置函数操作数据。

- batch(4)：按照顺序取出 4 行数据，最后一次输出可能小于 batch。
- repeat()：设置数据集重复执行指定的次数，在 batch 操作输出完毕后再执行此操作。如果在之前执行操作，相当于把整个数据集复制两次。为了配合输出次数，repeat()的参数一般默认为空。

在作者电脑中执行代码后会输出：

```
Epoch 1/3
782/782 [==============================] - 2s 2ms/step - loss: 0.3395 -
sparse_categorical_accuracy: 0.9036
Epoch 2/3
782/782 [==============================] - 2s 2ms/step - loss: 0.1614 -
sparse_categorical_accuracy: 0.9527
Epoch 3/3
782/782 [==============================] - 2s 2ms/step - loss: 0.1190 -
sparse_categorical_accuracy: 0.9648
Evaluate 评估:
157/157 [==============================] - 0s 2ms/step - loss: 0.1278 -
sparse_categorical_accuracy: 0.9633
{'loss': 0.12783484160900116,
 'sparse_categorical_accuracy': 0.9632999897003174}
```

另外需要注意，因为 tf.data 数据集会在每个周期结束时重置，所以可以在下一个周期中重复使用。如果只想在来自此数据集的特定数量批次上进行训练，则可以使用参数

steps_per_epoch，此参数可以指定在继续下一个周期之前，当前模型应该使用此数据集运行多少训练步骤。如果执行此操作，则不会在每个周期结束时重置数据集，而是会继续绘制接下来的批次，tf.data 数据集最终会用尽数据(除非它是无限循环的数据集)。例如在下面的实例中，演示了训练数据集中特定数量的批次的方法。

实例文件 xun09.py 的主要实现代码如下所示。

源码路径：daima\7\7-3\xun\xun09.py

```
model = get_compiled_model()

#准备训练数据集
train_dataset = tf.data.Dataset.from_tensor_slices((x_train, y_train))
train_dataset = train_dataset.shuffle(buffer_size=1024).batch(64)

#只训练每个 epoch 中的 100 个批次(即 64×100 个样本)
model.fit(train_dataset, epochs=3, steps_per_epoch=100)
```

在上述代码中，使用采纳数 steps_per_epoch 设置只训练每个 epoch 中的 100 个批次的样本。在作者电脑中执行代码后会输出：

```
Epoch 1/3
100/100 [==============================] - 0s 2ms/step - loss: 0.7755 -
sparse_categorical_accuracy: 0.7983
Epoch 2/3
100/100 [==============================] - 0s 2ms/step - loss: 0.3505 -
sparse_categorical_accuracy: 0.8981
Epoch 3/3
100/100 [==============================] - 0s 3ms/step - loss: 0.3197 -
sparse_categorical_accuracy: 0.9022
<tensorflow.python.keras.callbacks.History at 0x7f698e544198>
```

另外，还可以在函数 tf.keras.Model.fit()中将 tf.data 数据集的实例作为参数 validation_data 进行传递，这样 tf.data 数据集便当作验证数据集来处理。例如在下面的实例中，演示了将 tf.data 作为验证数据集进行训练的方法。

实例文件 xun10.py 的主要实现代码如下所示。

源码路径：daima\7\7-3\xun\xun10.py

```
model = get_compiled_model()

#准备训练数据集
train_dataset = tf.data.Dataset.from_tensor_slices((x_train, y_train))
train_dataset = train_dataset.shuffle(buffer_size=1024).batch(64)
```

```
# 准备验证数据集
val_dataset = tf.data.Dataset.from_tensor_slices((x_val, y_val))
val_dataset = val_dataset.batch(64)

model.fit(train_dataset, epochs=1, validation_data=val_dataset)
```

在上述代码中，当每个周期结束时，模型会迭代验证数据集并计算验证损失和验证指标。在作者电脑中执行代码后会输出：

```
782/782 [==============================] - 2s 3ms/step - loss: 0.3383 -
sparse_categorical_accuracy: 0.9050 - val_loss: 0.1783 -
val_sparse_categorical_accuracy: 0.9508
<tensorflow.python.keras.callbacks.History at 0x7f698e35b550>
```

如果只想对此数据集中的特定数量批次进行验证，则可以设置参数 validation_steps，此参数可以指定在中断验证并进入下一个周期之前，模型应使用验证数据集运行多少验证步骤。请看下面的实例，功能是通过参数 validation_steps 设置只使用数据集中的前 10 个 batch 批处理运行验证。

实例文件 xun11.py 的主要实现代码如下所示。

源码路径：daima\7\7-3\xun\xun11.py

```
#准备训练数据集
train_dataset = tf.data.Dataset.from_tensor_slices((x_train, y_train))
train_dataset = train_dataset.shuffle(buffer_size=1024).batch(64)

#准备验证数据集
val_dataset = tf.data.Dataset.from_tensor_slices((x_val, y_val))
val_dataset = val_dataset.batch(64)

model.fit(
    train_dataset,
    epochs=1,
    #通过参数“validation_steps”，设置只使用数据集中的前 10 个批处理运行验证
    validation_data=val_dataset,
    validation_steps=10,
)
```

验证会在当前 epoch 结束后进行，通过 validation_steps 设置验证使用的 batch 数量。假如 validation batch size=64(没必要和 train batch 相等)，而 validation_steps=100，steps 相当于 batch 数，则会从 validation data 中取 6400 个数据用于验证。如果在一次 step 后，在验证数据中剩下的数据足够下一次 step，则会继续从剩下的数据中选取，如果不够则会重新循环。在作者电脑中执行代码后会输出：

```
782/782 [==================] - 2s 2ms/step - loss: 0.3299 - sparse_categorical_
accuracy: 0.9067 - val_loss: 0.2966 - val_sparse_categorical_accuracy: 0.9250
<tensorflow.python.keras.callbacks.History at 0x7f698e35e400>
```

注意：当时用 Dataset 对象进行训练时，不能使用参数 validation_split(从训练数据生成预留集)，因为在使用 validation_split 功能时需要为数据集样本编制索引，而 Dataset API 通常无法做到这一点。

7.3.8 使用样本加权和类加权

在默认设置下，样本的权重由其在数据集中出现的频率决定。要想独立于样本频率来加权数据，可以通过如下两种方法实现。

1. 类权重

通过将字典传递给函数 Model.fit()中的参数 class_weight 的方式进行设置，此字典会将类索引映射到用于此类的样本的权重。另外，这也用在不重采样的情况下平衡类，或者用于更加重视特定类的模型训练。

请看下面的实例，这是一个 NumPy 例子，在其中使用类权重改变类 5(MNIST 数据集中的数字“5”)的权重。

实例文件 xun12.py 的主要实现代码如下所示。

源码路径：daima\7\7-3\xun\xun12.py

```
class_weight = {
    0: 1.0,
    1: 1.0,
    2: 1.0,
    3: 1.0,
    4: 1.0,
    # 为类"5"设置权重 2，确保类 2x 变得更重要
    5: 2.0,
    6: 1.0,
    7: 1.0,
    8: 1.0,
    9: 1.0,
}

print("分类权重")
model = get_compiled_model()
model.fit(x_train, y_train, class_weight=class_weight, batch_size=64, epochs=1)
```

在作者电脑中执行代码后会输出：

```
782/782 [==================] - 2s 2ms/step - loss: 0.3299 - sparse_categorical_
accuracy: 0.9067 - val_loss: 0.2966 - val_sparse_categorical_accuracy: 0.9250
<tensorflow.python.keras.callbacks.History at 0x7f698e35e400>
```

2. 样本权重

对于细粒度方面的控制，或者是不想构建分类器，则可以使用样本权重。样本权重数组是一个由数字组成的数组，这些数字用于设置每个样本在计算总损失时应当具有的权重。样本权重通常用于处理不平衡的分类问题，其理念是将更多的权重分配给罕见类。当使用的权重为 1 和 0 时，此数组可以作为损失函数的掩码(完全丢弃某些样本对总损失的贡献)。

当通过 NumPy 数据进行训练时，将参数 sample_weight 传递给函数 model.fit()来实现样本权重的功能。请看下面的实例，功能是通过 NumPy 数据训练时使用样本权重。

实例文件 xun13.py 的主要实现代码如下所示。

源码路径：daima\7\7-3\xun\xun13.py

```
sample_weight = np.ones(shape=(len(y_train),))
sample_weight[y_train == 5] = 2.0

print("Fit with sample weight")
model = get_compiled_model()
model.fit(x_train, y_train, sample_weight=sample_weight, batch_size=64, epochs=1)
```

在作者电脑中执行代码后会输出：

```
Fit with sample weight
782/782 [==============================] - 2s 2ms/step - loss: 0.3836 -
sparse_categorical_accuracy: 0.8981
<tensorflow.python.keras.callbacks.History at 0x7f9260373fd0>
```

当使用 tf.data 或任何其他类型的迭代器进行训练时，通过产生 (input_batch, label_batch, sample_weight_batch)元组的方式实现样本权重。请看下面的实例，功能是通过 Dataset 数据训练时使用样本权重。

实例文件 xun14.py 的主要实现代码如下所示。

源码路径：daima\7\7-3\xun\xun14.py

```
sample_weight = np.ones(shape=(len(y_train),))
sample_weight[y_train == 5] = 2.0
```

```
# 创建一个包含样本权重的数据集(返回元组中的第三个元素)
train_dataset = tf.data.Dataset.from_tensor_slices((x_train, y_train, sample_weight))

# Shuffle 洗牌和 slice 切片数据集
train_dataset = train_dataset.shuffle(buffer_size=1024).batch(64)

model = get_compiled_model()
model.fit(train_dataset, epochs=1)
```

在作者电脑中执行代码后会输出：

```
782/782 [==============================] - 2s 3ms/step - loss: 0.3661 -
sparse_categorical_accuracy: 0.9027
<tensorflow.python.keras.callbacks.History at 0x7f926015d898>
```

7.4 TensorFlow 图像视觉处理

经过本章前文内容的学习，已经了解了使用 TensorFlow 实现机器学习的基本知识。在本节的内容中，将进一步讲解使用 TensorFlow 实现图像视觉处理的知识。请看下面的实例文件 TFtu01.py，功能是训练一个神经网络模型，对运动鞋和衬衫等服装图像进行分类。本实例的目的是使用 tf.keras 来构建和训练模型的高级 API。

扫码观看本节视频讲解

源码路径：daima\7\7-4\TFtu01.pyy

7.4.1 导入需要的库

导入需要的库，打印输出当前的 TensorFlow 版本。

```
import tensorflow as tf
from tensorflow import keras

import numpy as np
import matplotlib.pyplot as plt

print(tf.__version__)
```

在笔者电脑中执行代码后会输出：

```
2.5.0
```

7.4.2 导入 Fashion MNIST 数据集

本实例使用了 Fashion MNIST 数据集，在该数据集中包含 10 个类别的 70000 张灰度图像。这些图像以低分辨率(28 像素×28 像素)展示了单件衣物、包和鞋子，如图 7-10 所示。

图 7-10 Fashion MNIST 数据集

推出数据集 Fashion MNIST 的目的是临时替代经典的 MNIST 数据集，后者常被用作计算机视觉机器学习程序的“Hello, World”。MNIST 数据集包含了手写数字(0、1、2 等)图像，其格式与在本实例中使用的衣物图像的格式相同。

在本实例中将使用 Fashion MNIST 数据集实现多样化处理，因为它比常规的 MNIST 数据集更具挑战性。这两个数据集都相对较小，都用于验证某个算法是否按预期工作。对于代码的测试和调试来说，它们都是很好的起点。在本实例中，将使用 60000 个图像来训练网络，使用 10000 个图像来评估网络学习对图像分类的准确率。我们可以直接从 TensorFlow 访问 Fashion MNIST。例如编写如下代码，直接从 TensorFlow 中导入和加载 Fashion MNIST 数据：

```
fashion_mnist = keras.datasets.fashion_mnist
(train_images, train_labels), (test_images, test_labels) = fashion_mnist.load_data()
```

在加载数据集时，会返回 4 个 NumPy 数组：

- train_images 和 train_labels 数组是训练集，即模型用于学习的数据。
- 测试集、test_images 和 test_labels 数组会被用来对模型进行测试。

图像是 28 像素×28 像素的 NumPy 数组，像素值介于 0 到 255 之间。标签是整数数组，介于 0 到 9 之间。这些标签对应于图像所代表的服装类，如表 7-1 所示。

表 7-1 标签对应图像所代表的服装类

标　签	类
0	T 恤/上衣
1	裤子
2	套头衫
3	连衣裙
4	外套
5	凉鞋
6	衬衫
7	运动鞋
8	包
9	短靴

在 Fashion MNIST 数据集中，每个图像都会被映射到一个标签。由于数据集不包括类的名称，所以将它们存储在图像的下方，以便在后面绘制图像时使用：

```
class_names = ['T-shirt/top', 'Trouser', 'Pullover', 'Dress', 'Coat',
               'Sandal', 'Shirt', 'Sneaker', 'Bag', 'Ankle boot']
```

7.4.3 浏览数据

在训练模型之前先浏览一下数据集的格式，通过以下代码可知在训练集中有 60000 个图像，每个图像由 28 像素×28 像素表示：

```
print(train_images.shapc)
```

执行后会输出：

```
(60000, 28, 28)
```

同样，在训练集中有 60000 个标签：

```
print(len(train_labels))
```

执行后会输出：

```
60000
```

每个标签都是一个 0 到 9 之间的整数：

```
print(train_labels)
```

执行后会输出：

```
array([9, 0, 0, ..., 3, 0, 5], dtype=uint8)
```

在测试集中有 10000 个图像，每个图像都由 28 像素×28 像素表示：

```
print(test_images.shape)
```

执行后会输出：

```
(10000, 28, 28)
```

在测试集中包含 10000 个图像标签：

```
print(len(test_labels))
```

执行后会输出：

```
10000
```

7.4.4 预处理数据

在训练神经网络之前，必须对数据进行预处理。如果通过如下代码检查训练集中的第一个图像，会看到像素值处于 0 到 255 之间：

```
plt.figure()
plt.imshow(train_images[0])
plt.colorbar()
plt.grid(False)
plt.show()
```

执行上述代码后的效果如图 7-11 所示。

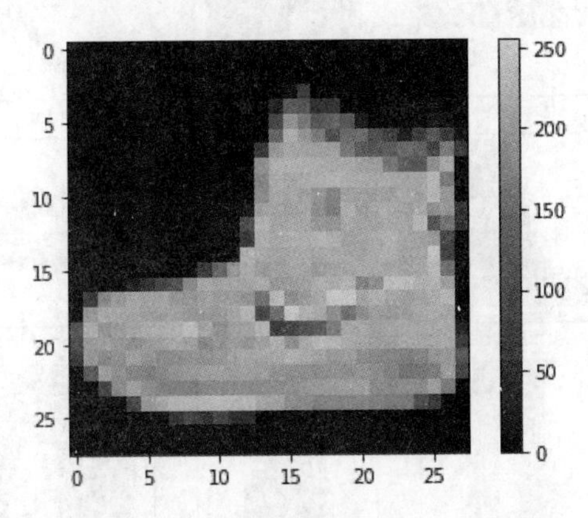

图 7-11 检查训练集中的第一个图像

将这些值缩小至 0 到 1 之间，然后将其反馈到神经网络模型。为此，需要通过如下代码将这些值除以 255，并必须以相同的方式对训练集和测试集进行预处理：

```
train_images = train_images / 255.0
test_images = test_images / 255.0
```

为了验证数据的格式是否正确，以及是否已准备好构建和训练网络，通过如下代码显示训练集中的前 25 个图像，并在每个图像下方显示类名称：

```
plt.figure(figsize=(10,10))
for i in range(25):
   plt.subplot(5,5,i+1)
   plt.xticks([])
   plt.yticks([])
   plt.grid(False)
   plt.imshow(train_images[i], cmap=plt.cm.binary)
   plt.xlabel(class_names[train_labels[i]])
plt.show()
```

运行上述代码后会显示训练集中的前 25 个图像，如图 7-12 所示。

图 7-12 训练集中的前 25 个图像

7.4.5 构建模型

构建神经网络需要先配置模型的层，然后再编译模型。接下来首先设置层。神经网络的基本组成部分是层，层会从向其馈送的数据中提取表示形式，希望这些表示形式有助于解决手头上的问题。下面是本实例创建层的代码：

```
model = keras.Sequential([
    keras.layers.Flatten(input_shape=(28, 28)),
    keras.layers.Dense(128, activation='relu'),
    keras.layers.Dense(10)
])
```

通过上述代码，在网络中的第一层 tf.keras.layers.Flatten 将图像格式从二维数组(28 × 28 像素)转换成一维数组(28 × 28 = 784 像素)，将该层视为图像中未堆叠的像素行并将其排列起来。在该层没有要学习的参数，它只会重新格式化数据。

在展平像素后，网络会包括两个 tf.keras.layers.Dense 层的序列，它们是密集连接或全连接神经层。第一个 Dense 层有 128 个节点(或神经元)，第二个(也是最后一个)层会返回一个长度为 10 的 logits 数组。每个节点都包含一个得分，用来表示当前图像属于 10 个类中的哪一类。

7.4.6 编译模型

在准备对模型进行训练之前，还需要再对其进行一些设置。以下内容是在模型的编译步骤中添加的。

- 损失函数：用于测量模型在训练期间的准确率。一般会希望最小化，以便将模型“引导”到正确的方向上。
- 优化器：决定模型如何根据其看到的数据和自身的损失函数进行更新。
- 指标：用于监控训练和测试步骤。

以下实例使用了准确率，即被正确分类的图像的比率。下面是本实例的模型编译代码：

```
model.compile(optimizer='adam',

loss=tf.keras.losses.SparseCategoricalCrossentropy(from_logits=True),
          metrics=['accuracy'])
```

7.4.7 训练模型

在 TensorFlow 中，训练神经网络模型的基本步骤如下。

- 将训练数据馈送给模型。在本例中，训练数据位于 train_images 和 train_labels 数组中。
- 模型学习将图像和标签关联起来。
- 要求模型对测试集(在本例中为 test_images 数组)进行预测。
- 验证预测是否与 test_labels 数组中的标签相匹配。

(1) 向模型馈送数据。

在训练时需要调用方法 model.fit()，这样命名的原因是因为该方法会将模型与训练数据进行“拟合”。

```
model.fit(train_images, train_labels, epochs=10)
```

执行上述代码后会打印输出训练过程：

```
Epoch 1/10
1875/1875 [==============================] - 3s 1ms/step - loss: 0.4924 -
accuracy: 0.8265
Epoch 2/10
1875/1875 [==============================] - 3s 1ms/step - loss: 0.3698 -
accuracy: 0.8669
Epoch 3/10
1875/1875 [==============================] - 3s 1ms/step - loss: 0.3340 -
accuracy: 0.8781
Epoch 4/10
1875/1875 [==============================] - 3s 1ms/step - loss: 0.3110 -
accuracy: 0.8863
Epoch 5/10
1875/1875 [==============================] - 3s 1ms/step - loss: 0.2924 -
accuracy: 0.8936
Epoch 6/10
1875/1875 [==============================] - 3s 1ms/step - loss: 0.2776 -
accuracy: 0.8972
Epoch 7/10
1875/1875 [==============================] - 3s 1ms/step - loss: 0.2659 -
accuracy: 0.9021
Epoch 8/10
1875/1875 [==============================] - 3s 1ms/step - loss: 0.2543 -
accuracy: 0.9052
Epoch 9/10
1875/1875 [==============================] - 3s 1ms/step - loss: 0.2453 -
accuracy: 0.9084
Epoch 10/10
1875/1875 [==============================] - 3s 1ms/step - loss: 0.2366 -
accuracy: 0.9122
<tensorflow.python.keras.callbacks.History at 0x7fc85fa4f2e8>
```

在训练模型期间，会输出显示损失和准确率指标。通过上述执行结果可以看出，此模型在训练数据上的准确率达到了 0.91(或 91%)左右。

(2) 评估准确率。

接下来，编写如下代码比较模型在测试数据集上的表现：

```
test_loss, test_acc = model.evaluate(test_images,  test_labels, verbose=2)

print('\nTest accuracy:', test_acc)
```

执行代码后会输出：

```
313/313 - 0s - loss: 0.3726 - accuracy: 0.8635

Test accuracy: 0.8634999990463257
```

上述执行结果表明，模型在测试数据集上的准确率略低于训练数据集。训练准确率和测试准确率之间的差距代表过拟合。过拟合是指机器学习模型在新的、以前未曾见过的输入上的表现不如在训练数据上的表现。过拟合的模型会“记住”训练数据集中的噪声和细节，从而对模型在新数据上的表现产生负面影响。

(3) 预测。

在模型经过训练后，可以使用这个模型对一些图像进行预测，此时模型具有线性输出功能，即 logits。我们可以在上面附加一个 softmax 层，将 logits 转换成更容易理解的概率，然后使用模型预测测试集中每个图像的标签：

```
probability_model = tf.keras.Sequential([model,
                                         tf.keras.layers.Softmax()])

predictions = probability_model.predict(test_images)
predictions[0]
```

我们来看看第一个预测结果：

```
array([6.9982241e-07, 5.5403369e-08, 1.8353174e-07, 1.4761626e-07,
       2.4380807e-07, 1.9273469e-04, 1.8122660e-06, 6.5027133e-02,
       1.7891599e-06, 9.3477517e-01], dtype=float32)
```

上述预测结果是一个包含 10 个数字的数组，分别代表模型对 10 种不同服装中每种服装的“置信度”。我们可以通过如下代码查看哪个标签的置信度值最大：

```
np.argmax(predictions[0])
```

执行代码后会输出：

```
9
```

通过上述执行效果可知，该模型非常确信这个图像是短靴，或 class_names[9]。通过检查测试标签发现这个分类是正确的：

```
print(test_labels[0])
```

执行代码后会输出：

```
9
```

我们可以编写如下代码将上述过程绘制成图表，看看使用模型预测全部 10 个类的结果：

```
def plot_image(i, predictions_array, true_label, img):
  predictions_array, true_label, img = predictions_array, true_label[i], img[i]
  plt.grid(False)
  plt.xticks([])
  plt.yticks([])

  plt.imshow(img, cmap=plt.cm.binary)

  predicted_label = np.argmax(predictions_array)
  if predicted_label == true_label:
    color = 'blue'
  else:
    color = 'red'

  plt.xlabel("{} {:2.0f}% ({})".format(class_names[predicted_label],
                                100*np.max(predictions_array),
                                class_names[true_label]),
                                color=color)

def plot_value_array(i, predictions_array, true_label):
  predictions_array, true_label = predictions_array, true_label[i]
  plt.grid(False)
  plt.xticks(range(10))
  plt.yticks([])
  thisplot = plt.bar(range(10), predictions_array, color="#777777")
  plt.ylim([0, 1])
  predicted_label = np.argmax(predictions_array)

  thisplot[predicted_label].set_color('red')
  thisplot[true_label].set_color('blue')
```

(4) 验证预测结果。

在模型经过训练后，可以使用这个模型对一些图像进行预测。我们首先看第 0 个图像、预测结果和预测数组。正确的预测标签为蓝色，错误的预测标签为红色，数字表示预测标签的百分比(总计为 100)。代码如下：

```
i = 0
plt.figure(figsize=(6,3))
plt.subplot(1,2,1)
plot_image(i, predictions[i], test_labels, test_images)
plt.subplot(1,2,2)
plot_value_array(i, predictions[i],  test_labels)
plt.show()
```

执行代码后会在绘制的图表中显示对第 0 个图像的预测结果，如图 7-13 所示。

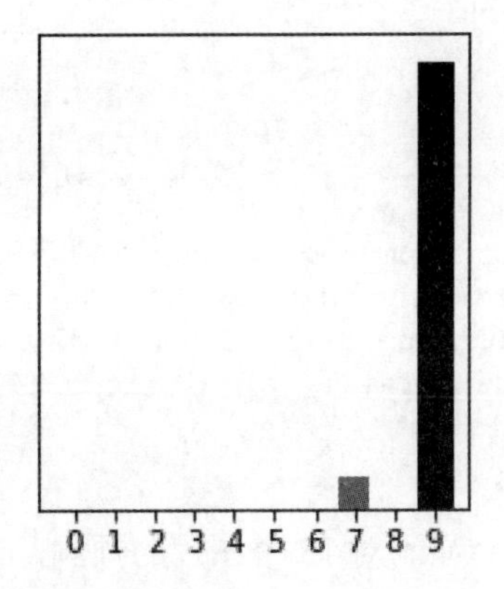

图 7-13 对第 0 个图像的预测结果

也可以通过如下代码看第 12 个图像的预测结果：

```
i = 12
plt.figure(figsize=(6,3))
plt.subplot(1,2,1)
plot_image(i, predictions[i], test_labels, test_images)
plt.subplot(1,2,2)
plot_value_array(i, predictions[i],  test_labels)
plt.show()
```

执行代码后会显示对第 12 个图像的预测结果，如图 7-14 所示。

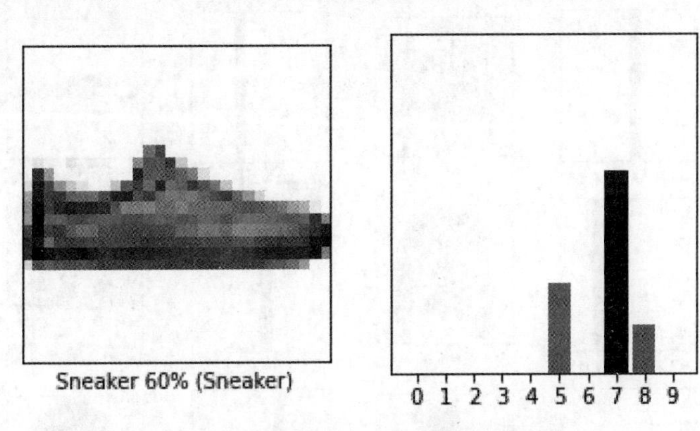

图 7-14 对第 12 个图像的预测结果

注意，如果使用模型预测绘制的几张图像，即使预测率很高，模型也可能出错。请看下面的代码，绘制一个测试图像和对应的预测标签及真实标签，其中用蓝色显示正确预测，用红色显示错误预测：

```
num_rows = 5
num_cols = 3
num_images = num_rows*num_cols
plt.figure(figsize=(2*2*num_cols, 2*num_rows))
for i in range(num_images):
  plt.subplot(num_rows, 2*num_cols, 2*i+1)
  plot_image(i, predictions[i], test_labels, test_images)
  plt.subplot(num_rows, 2*num_cols, 2*i+2)
  plot_value_array(i, predictions[i], test_labels)
plt.tight_layout()
plt.show()
```

执行代码后会显示对 15 个图片的预测，效果如图 7-15 所示。

图 7-15　对 15 个图片的预测

7.4.8 使用训练好的模型

最后编写如下所示的代码，使用前面训练好的模型对单个图像进行预测：

```
#从测试数据集中获取图像
img = test_images[1]

print(img.shape)
```

执行代码后会输出：

```
(28, 28)
```

tf.keras 模型经过了优化，可以同时对一个批数据或一组样本进行预测。因此，即便我们只使用一个图像，也需要将其添加到列表中：

```
#将图像添加到一个批处理中，它是唯一的成员。
img = (np.expand_dims(img,0))

print(img.shape)
```

执行代码后会输出：

```
(1, 28, 28)
```

现在通过如下代码预测这个图像的正确标签：

```
predictions_single = probability_model.predict(img)

print(predictions_single)
```

执行代码后会输出：

```
[[1.0675135e-05 2.4023437e-12 9.9772269e-01 1.3299730e-09 1.2968916e-03
  8.7469149e-14 9.6970733e-04 5.4669354e-19 2.4514609e-11 1.8405429e-12]]
```

通过如下代码绘制预测结果：

```
plot_value_array(1, predictions_single[0], test_labels)
_ = plt.xticks(range(10), class_names, rotation=45)
```

执行代码后的效果如图 7-16 所示。

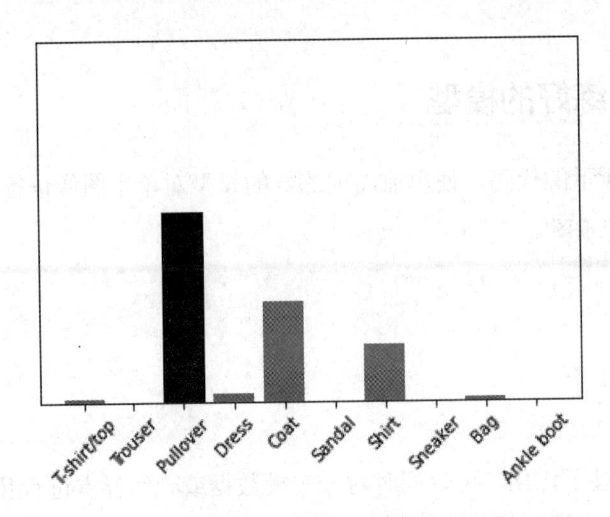

图 7-16 绘制的预测结果

在 TensorFlow 中，keras.Model.predict 会返回一组列表，每个列表对应一批数据中的每个图像。例如通过如下代码，可以在批次中获取对(唯一)图像的预测：

```
print(np.argmax(predictions_single[0]))
```

执行代码后会输出：

```
2
```

第8章 国内常用的第三方人脸识别平台

在本书前面的内容中，讲解了在 Python 程序中使用各种库实现图像处理和人脸识别的知识。为了进一步提高开发效率，国内一线开发公司推出了自己的在线识别 API，开发者可以调用它们的 API 实现人脸识别功能。在本章的内容中，将详细讲解国内常用的第三方人脸识别平台的知识，为读者步入本书后面知识的学习打下基础。

8.1 百度 AI 开放平台

百度 AI 开放平台为开发者提供了全球领先的语音、图像、NLP 等多项人工智能技术，开放了对话式人工智能系统、智能驾驶系统两大行业生态，以及共享 AI 领域最新的应用场景和解决方案。

扫码观看本节视频讲解

8.1.1 百度 AI 开放平台介绍

在百度 AI 平台中，提供了人脸实名认证、短语音识别、人脸离线识别、文字识别、证件识别、语音合成、地址识别和车牌识别等功能，如图 8-1 所示。

图 8-1 百度 AI

8.1.2 使用百度 AI 之前的准备工作

为了提高开发效率，降低开发成本，我们使用百度公司提供的 AI 接口实现在线人脸识别功能，具体原理如下。

- 向百度 AI 发送人脸检测请求，让百度 AI 去完成人脸识别，百度 AI 返回识别结果。
- 发送请求不是任意的网络请求都能够接受，必须有百度提供的访问令牌(access_token)。

1. 发送请求前的准备工作

在向百度 AI 发送人脸检测请求之前，必须先获取 access_token，并注册人脸识别 API 的如下参数。

- client_id：当前应用程序的标识。
- client_secret：决定是否有访问权限。

也就是说，开发者需要注册百度 API 获取 id 与 secret，在注册时使用百度账号进行注册。注册后需要创建人脸识别应用，在创建应用后才会获得 id 与 secret：

- id：应用的 API Key，例如 kSD6zWfxpki2AKWtysCUe0nS。
- secret：应用的 Secret Key，例如 uNXjdRa7SbYwt0EgBdRwsmQYX6VADGx8。

在使用 requests.get(host)发送请求后，最终得到字典数据格式的数据，从字典中取出键为 access_token 的值即可得到 access_token 的值。

2. 开始发送请求

发送请求，通过网络请求方式让百度 AI 进行人脸识别。让百度 AI 检测一张画面(图片)是否存在人脸以及人脸的一些属性。通过函数 requests.post()完成识别请求，返回检测到的结果。返回的结果数据是一个字典，在里面保存了多项数据内容，通过键值对进行表示。

3. 完成人脸搜索

在百度 AI 库中搜索是否存在对应的人脸，有则实现签到功能。

综上所述，我们在使用百度 AI 实现人脸识别功能之前，需要先创建一个百度 AI 应用程序并获得 access_token，具体流程如下。

(1) 输入网址 https://ai.baidu.com/登录百度 AI 主页，如图 8-2 所示。

(2) 单击顶部导航中的“开放能力”链接，然后在弹出的子链接中依次单击“人脸与人体识别”“人脸识别”，如图 8-3 所示。

(3) 在弹出的新界面中单击“立即使用”按钮，如图 8-4 所示。这一步需要输入百度账号登录百度智能云，如果没有百度账号则需要先申请一个。

图 8-2 百度 AI 主页

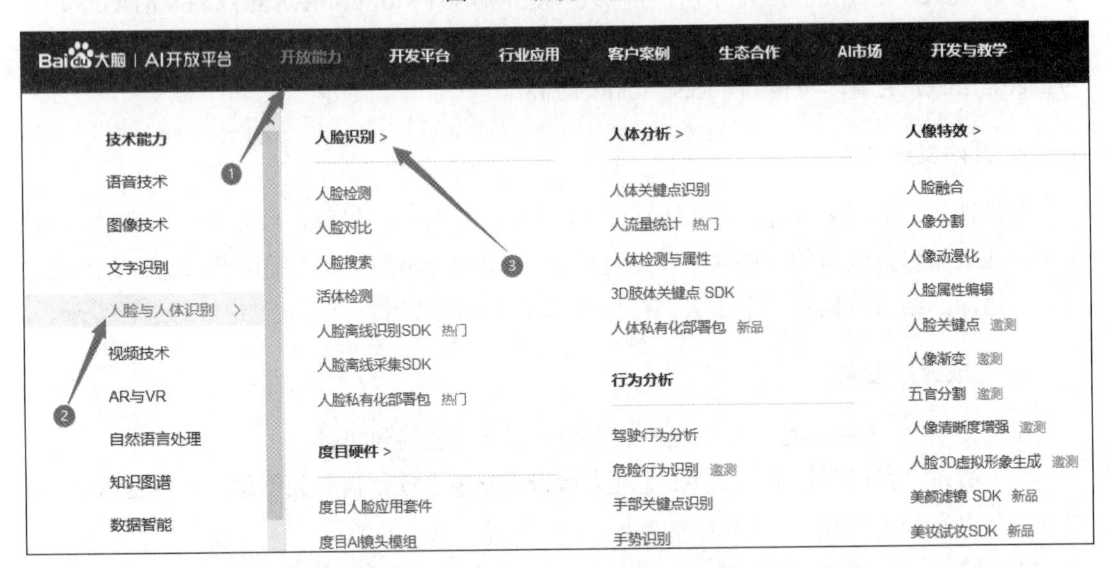

图 8-3 依次单击“开放能力”“人脸与人体识别”“人脸识别”

(4) 在弹出的新界面中依次单击“人脸实名认证”“创建应用”链接，如图 8-5 所示。

(5) 在弹出的新界面中设置应用程序的“简称”和“描述信息”，例如都填写“人脸检测”，其他选项都是默认，最后单击下面的“立即创建”按钮，如图 8-6 所示。

(6) 创建成功后，在应用列表中会显示刚刚创建的应用，并可以查看这个应用的 API

Key 和 Secret Key，如图 8-7 所示。通过使用 API Key 和 Secret Key 可以获取需要用到的 access_token。

图 8-4　单击“立即使用”按钮

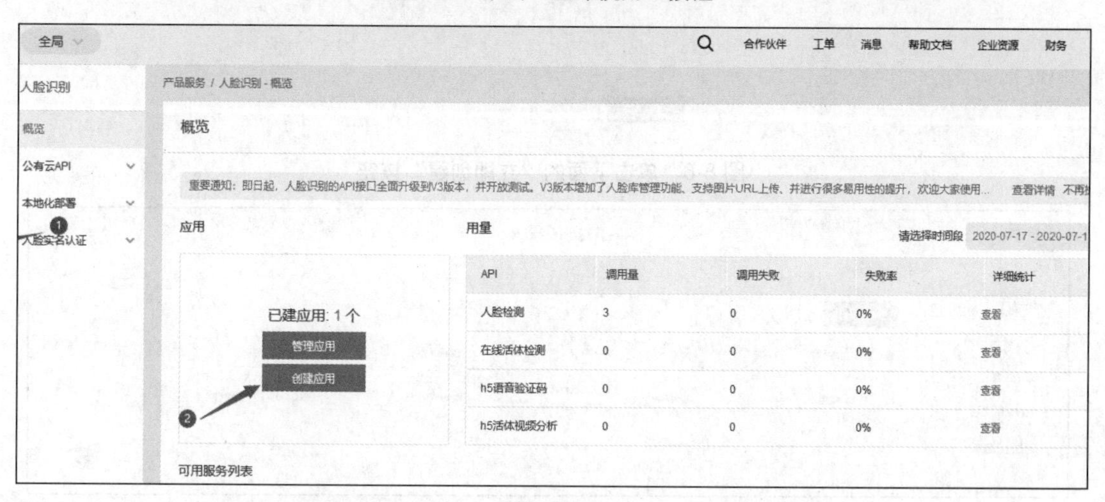

图 8-5　依次单击“人脸实名认证”“创建应用”链接

图 8-6　单击下面的“立即创建”按钮

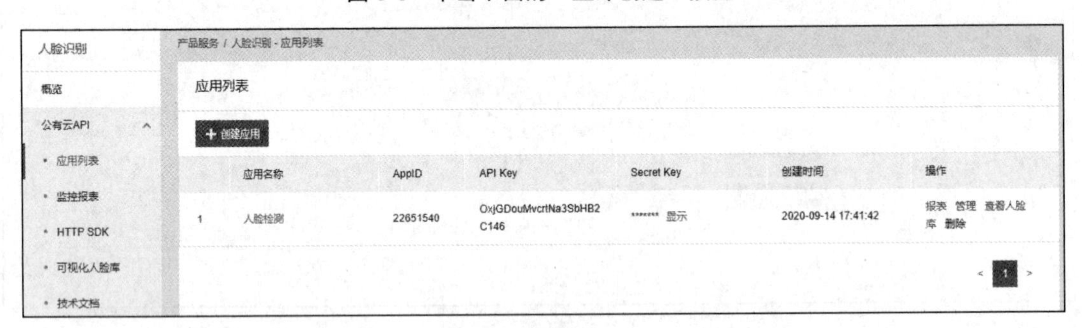

图 8-7　刚刚创建的应用

8.1.3 基于百度 AI 平台的人脸识别

请看下面的实例文件 baidu.py，功能是调用百度 AI 检测某张照片的基本信息，返回这张照片中人物的年龄和颜值信息。

源码路径：daima\8\8-1\baidu.py

```
from aip import AipFace
import base64

APP_ID = '22651540'
API_KEY = ''
SECRET_KEY = ''

client=AipFace(APP_ID,API_KEY,SECRET_KEY)#print(client)#读取照片文件
file=open("111.jpg","rb")
#对照片二进制进行 base64 编码
img=base64.b64encode(file.read()).decode()
print(img)                                          #检测 img 这个照片中的人脸信息
options={"max_face_num":10,                         #最多的人脸数量
         "face_field":"age,beauty"                  #希望检测结果数据中包含年龄，颜值信息
         }
data=client.detect(img,"BASE64",options)
print(data)
if data["error_code"]==0:                               #返回数据中的检测结果
  result=data["result"]                                #人脸数量
  faceNum=result["face_num"]
  faceList=result["face_list"]
  for face in faceList:
      print(f"年龄：{face['age']}岁，颜值：{face['beauty']}")
else:
  print("出错了")
```

执行代码后会输出照片的信息：

```
{'error_code': 0, 'error_msg': 'SUCCESS', 'log_id': 6594151535750, 'timestamp':
1623674390, 'cached': 0, 'result': {'face_num': 1, 'face_list': [{'face_token':
'3132f502fd35b0b632a613414d7701ef', 'location': {'left': 37.77, 'top': 260.84,
'width': 211, 'height': 190, 'rotation': 0}, 'face_probability': 1, 'angle':
{'yaw': -4.76, 'pitch': 9.01, 'roll': 1.28}, 'age': 2, 'beauty': 48.25}]}}
年龄：2 岁，颜值：48.25
```

再看下面的实例文件 baidu02.py，功能是开发一个 tkinter 桌面程序，弹出文件选择框供

用户选择一幅照片，然后调用百度 AI 返回这张照片中人物的年龄、性别、是否戴眼镜、人种和颜值信息。

源码路径：daima\8\8-1\baidu02.py

```
import requests
import base64
import tkinter.filedialog

def get_access_token(client_id, client_secret):
    # client_id 为官网获取的AK， client_secret 为官网获取的SK
    # 帮助文档
    # https://ai.baidu.com/docs#/Auth/top
    # 帮助文档中python 代码基于python2,本文已经转换为python3x 调试通过。
    host = 'https://aip.baidubce.com/oauth/2.0/token?grant_type=client_
        credentials&client_id=' + client_id + '&client_secret=' + client_secret
    header = {'Content-Type': 'application/json; charset=UTF-8'}
    response1 = requests.post(url=host, headers=header) # <class
              'requests.models.Response'>
    json1 = response1.json()  # <class 'dict'>
    access_token = json1['access_token']

    return access_token

def open_pic2base64():
    # 本地图片地址，根据自己的图片进行修改
    # 打开本地图片，并转化为base64
    root = tkinter.Tk()  # 创建一个Tkinter.Tk()实例
    root.withdraw()  # 将Tkinter.Tk()实例隐藏
    file_path = tkinter.filedialog.askopenfilename(title=u'选择文件')
    f = open(file_path, 'rb')
    img = base64.b64encode(f.read()).decode('utf-8')
    return img


def bd_rec_face(client_id, client_secret):
    # 识别人脸，给出性别、年龄、人种、颜值分数、是否带眼镜等信息
    # 帮助文档中python 代码基于python2,本文已经转换为python3x 调试通过

    request_url = "https://aip.baidubce.com/rest/2.0/face/v3/detect"
    params = {"image": open_pic2base64(), "image_type": "BASE64",
           "face_field": "age,beauty,glasses,gender,race"}
    header = {'Content-Type': 'application/json'}

    access_token = get_access_token(client_id, client_secret)
                  # '[调用鉴权接口获取的token]'
```

```
    request_url = request_url + "?access_token=" + access_token

    request_url = request_url + "?access_token=" + access_token
    response1 = requests.post(url=request_url, data=params, headers=header)
    json1 = response1.json()
    print("性别为", json1["result"]["face_list"][0]['gender']['type'])
    print("年龄为", json1["result"]["face_list"][0]['age'], '岁')
    print("人种为", json1["result"]["face_list"][0]['race']['type'])
    print("颜值评分为", json1["result"]["face_list"][0]['beauty'], '分/100 分')
    print("是否带眼镜", json1["result"]["face_list"][0]['glasses']['type'])

if __name__ == '__main__':
    # 以下为代码功能测试:
    # 账户 id, client_id 为官网获取的 AK, client_secret 为官网获取的 SK。
    # https://console.bce.baidu.com/ai/?fromai=1#/ai/face/app/list
    APP_ID = '22651540'
    API_KEY = ''
    SECRET_KEY = ''

    # 实例 1: 人脸识别
    bd_rec_face(API_KEY, SECRET_KEY )
```

例如选择识别某幅照片后会输出：

```
性别为 male
年龄为 35 岁
人种为 yellow
颜值评分为 39.99 分/100 分
是否带眼镜 common
```

再看下面的实例文件 baidu03.py，功能是调用百度 AI 识别出某张照片中的人脸，如果是男性则用红色矩形框标记出人脸，如果是女性则用绿色矩形框标记出人脸。

源码路径：daima\8\8-1\baidu03.py

```
# 导入 base64 库
import base64
# 导入百度 AI 人脸识别库
from aip import AipFace
# 导入 Pillow 库
from PIL import Image,ImageDraw

# 定义常量
APP_ID = ' '
API_KEY = ' '
SECRET_KEY = ' '
```

```
# 初始化 AipFace 对象
client = AipFace(APP_ID, API_KEY, SECRET_KEY)

def baiduAiFace(filename:str):
    # 打开图像
    f = open(filename, 'rb')
    # 对图像进行 base64 编码
    base64_data = base64.b64encode(f.read())
    # 对原图像进行 base64 解码,得到所处理图像
    image = base64_data.decode()
    # 定义图像类型
    imageType = "BASE64"
    # 定义可选参数
    options = {}
    # 定义可识别的人脸数为最大值 10
    options['max_face_num'] = 10
    # 定义可选返回信息，性别 gender
    options['face_field'] = 'gender'
    # 调用 API 接口,返回值类型为字典
    imageDic = client.detect(image,imageType,options=options)
    # 获取人脸信息列表
    face_list = imageDic['result']["face_list"]
    # 由人脸信息列表中提取所需性别、位置信息
    for item in face_list:
        gender = item['gender']['type']
        left = item['location']['left']
        top = item['location']['top']
        width = item['location']['width']
        height = item['location']['height']
        # 构建迭代器
        yield (gender,left,top,width,height)

def DrawImage(filename:str):
    # 打开图像
    image = Image.open(filename)
    # 生成一个可用于画图的对象
    draw = ImageDraw.Draw(image)
    # 使用百度平台返回的人脸坐标信息在图像中人脸的位置画一个矩形框
    for _,item in enumerate(baiduAiFace(filename)):
        # 用 gender,left,top,width,height 分别表示性别、人脸框左上角的横、纵坐标以及
        # 人脸框的宽度、高度
        (gender,left,top,width,height) = item
        # 根据性别定义矩形框的颜色,男性为红色(255,0,0),女性为绿色(0,255,0)
        outRGB = (255,0,0) if gender == 'male' else (0,255,0)
```

```
        draw.polygon([(left,top),(left+width,top),(left+width,height+top),
                    (left,height+top)], outline=outRGB)
    # 展示图像
    image.show()

if __name__ == '__main__':
DrawImage('111.jpg')
```

执行代码后会用矩形框标记处图片 111.jpg 中的人脸，如图 8-8 所示。

图 8-8 执行效果

8.2 科大讯飞 AI 开放平台

科大讯飞成立于 1999 年 12 月 30 日，专业从事智能语音及语言技术研究、软件及芯片产品开发、语音信息服务及电子政务系统集成，拥有灵犀语音助手、讯飞输入法等优秀产品。科大讯飞为开发者和客户提供了一整套的 AI 平台，可以快速实现 AI 相关功能。

扫码观看本节视频讲解

8.2.1 科大讯飞 AI 开放平台介绍

在科大讯飞 AI 平台中，提供了语音处理、图像识别、自然语言处理、人脸识别、文字识别、医疗服务等功能，如图 8-9 所示。

图 8-9　科大讯飞 AI

8.2.2　申请试用

和百度在线 AI 一样，开发者需要在线获取科大讯飞的 API Key 和 API 密钥，才可以在程序中使用科大讯飞的 AI 功能。对于普通开发者来说，可以在科大讯飞网站申请免费试用功能，通过实名认证后可以创建一个应用，如图 8-10 所示。

图 8-10　创建一个应用

在创建一个应用后会自动生成 3 个参数：Appid、APIKey 和 APISecret，将这 3 个参数添加到自己的 Python 程序中即可调用科大讯飞的在线 AI 功能。

8.2.3 基于科大讯飞 AI 的人脸识别

请看下面的实例文件 face_compare_keda.py，功能是基于科大讯飞自研的人脸算法，对比两张照片中的人脸信息，判断是否是同一个人并返回相似度得分。该功能是通过 HTTP API 的方式给开发者提供一个通用的接口。HTTP API 适用于一次性交互数据传输的 AI 服务场景——块式传输。在程序中需要先填写在科大讯飞官方网站创建的应用的 Appid、APIKey 和 APISecret。

源码路径：daima\8\8-2\face_compare_keda.py

```
class AssembleHeaderException(Exception):
    def __init__(self, msg):
        self.message = msg

class Url:
    def __init__(this, host, path, schema):
        this.host = host
        this.path = path
        this.schema = schema
        pass


# 进行 sha256 加密和 base64 编码
def sha256base64(data):
    sha256 = hashlib.sha256()
    sha256.update(data)
    digest = base64.b64encode(sha256.digest()).decode(encoding='utf-8')
    return digest


def parse_url(requset_url):
    stidx = requset_url.index("://")
    host = requset_url[stidx + 3:]
    schema = requset_url[:stidx + 3]
    edidx = host.index("/")
    if edidx <= 0:
        raise AssembleHeaderException("invalid request url:" + requset_url)
    path = host[edidx:]
    host = host[:edidx]
    u = Url(host, path, schema)
    return u


def assemble_ws_auth_url(requset_url, method="GET", api_key="",
api_secret=""):
```

```
    u = parse_url(requset_url)
    host = u.host
    path = u.path
    now = datetime.now()
    date = format_date_time(mktime(now.timetuple()))
    print(date)
    # date = "Thu, 12 Dec 2019 01:57:27 GMT"
    signature_origin = "host: {}\ndate: {}\n{} {} HTTP/1.1".format(host, date,
                        method, path)
    print(signature_origin)
    signature_sha = hmac.new(api_secret.encode('utf-8'), signature_
                origin.encode('utf-8'), digestmod=hashlib.sha256).digest()
    signature_sha = base64.b64encode(signature_sha).decode(encoding='utf-8')
    authorization_origin = "api_key=\"%s\", algorithm=\"%s\", headers=\"%s\",
                            signature=\"%s\"" % (
        api_key, "hmac-sha256", "host date request-line", signature_sha)
    authorization = base64.b64encode(authorization_origin.encode
                    ('utf-8')).decode(encoding='utf-8')
    print(authorization_origin)
    values = {
        "host": host,
        "date": date,
        "authorization": authorization
    }

    return requset_url + "?" + urlencode(values)

def gen_body(appid, img1_path, img2_path, server_id):
    with open(img1_path, 'rb') as f:
        img1_data = f.read()
    with open(img2_path, 'rb') as f:
        img2_data = f.read()
    body = {
        "header": {
            "app_id": appid,
            "status": 3
        },
        "parameter": {
            server_id: {
                "service_kind": "face_compare",
                "face_compare_result": {
                    "encoding": "utf8",
                    "compress": "raw",
                    "format": "json"
                }
            }
        },
        "payload": {
            "input1": {
```

```
                "encoding": "jpg",
                "status": 3,
                "image": str(base64.b64encode(img1_data), 'utf-8')
            },
            "input2": {
                "encoding": "jpg",
                "status": 3,
                "image": str(base64.b64encode(img2_data), 'utf-8')
            }
        }
    }
    return json.dumps(body)


def run(appid, apikey, apisecret, img1_path, img2_path, server_id='s67c9c78c'):
    url = 'http://api.xf-yun.com/v1/private/{}'.format(server_id)
    request_url = assemble_ws_auth_url(url, "POST", apikey, apisecret)
    headers = {'content-type': "application/json", 'host': 'api.xf-yun.com',
              'app_id': appid}
    print(request_url)
    response = requests.post(request_url, data=gen_body(appid, img1_path,
                                 img2_path, server_id), headers=headers)
    resp_data = json.loads(response.content.decode('utf-8'))
    print(resp_data)
    print(base64.b64decode(resp_data['payload']['face_compare_result']
          ['text']).decode())

#请填写控制台获取的APPID、APISecret、APIKey以及要比对的图片路径
if __name__ == '__main__':
    run(
        appid='11b6f6bb',
        apisecret='    ',
        apikey='    ',
        img1_path=r'111.jpg',
        img2_path=r'222.jpg',
    )
```

执行代码后会调用科大讯飞的在线 AI 对比两张照片 111.jpg 和 222.jpg，返回如下对比信息：

```
Tue, 15 Jun 2021 07:09:40 GMT
host: api.xf-yun.com
date: Tue, 15 Jun 2021 07:09:40 GMT
POST /v1/private/s67c9c78c HTTP/1.1
api_key="431ee6a26215785be1c4c75c61178a06", algorithm="hmac-sha256",
headers="host date request-line",
signature="gN03v/tVvUfvRimPER0QXPpgPDja8buQ3VPrnBNVSAc="
http://api.xf-yun.com/v1/private/s67c9c78c?host=api.xf-yun.com&date=Tue%2C+
15+Jun+2021+07%3A09%3A40+GMT&authorization=YXBpX2tleT0iNDMxZWU2YTI2MjE1Nzg1
```

```
YmUxYzRjNzVjNjExNzhhMDYiLCBhbGdvcml0aG09ImhtYWMtc2hhMjU2IiwgaGVhZGVycz0iaG9
zdCBkYXRlIHJlcXVlc3QtbGluZSIsIHNpZ25hdHVyZT0iZ04wM3YvdFZ2VWZ2UmltUEVSMFFYUH
BnUERqYThidVEzVlBybkJOVlNBYz0i
{'header': {'code': 0, 'message': 'success', 'sid':
'ase000d9780@hu17a0e810a0a0210882'}, 'payload': {'face_compare_result':
{'compress': 'raw', 'encoding': 'utf8', 'format': 'json', 'text':
'ewoJInJldCIgOiAwLAoJInNjb3JlIiA6IDAuODExNjIwNDczODYxNjk0MzQKfQo='}}}
{
    "ret" : 0,
    "score" : 0.81162047386169434
}
```


再看下面的实例文件face_detect_keda.py，功能是基于科大讯飞自研的人脸算法，对指定图片中的人脸进行精准定位并标记，分析人脸性别、表情、口罩等属性信息。在程序中需要先填写在科大讯飞官方网站创建的应用的Appid、APIKey和APISecret。

源码路径：daima\8\8-2\face_detect_keda.py

```python
class AssembleHeaderException(Exception):
    def __init__(self, msg):
        self.message = msg

class Url:
    def __init__(this, host, path, schema):
        this.host = host
        this.path = path
        this.schema = schema
        pass

# 进行sha256加密和base64编码
def sha256base64(data):
    sha256 = hashlib.sha256()
    sha256.update(data)
    digest = base64.b64encode(sha256.digest()).decode(encoding='utf-8')
    return digest

def parse_url(requset_url):
    stidx = requset_url.index("://")
    host = requset_url[stidx + 3:]
    schema = requset_url[:stidx + 3]
    edidx = host.index("/")
    if edidx <= 0:
        raise AssembleHeaderException("invalid request url:" + requset_url)
    path = host[edidx:]
    host = host[:edidx]
    u = Url(host, path, schema)
    return u
```

```
def assemble_ws_auth_url(requset_url, method="GET", api_key="", api_secret=""):
    u = parse_url(requset_url)
    host = u.host
    path = u.path
    now = datetime.now()
    date = format_date_time(mktime(now.timetuple()))
    print(date)
    # date = "Thu, 12 Dec 2019 01:57:27 GMT"
    signature_origin = "host: {}\ndate: {}\n{} {} HTTP/1.1".format(host, date,
                        method, path)
    print(signature_origin)
    signature_sha = hmac.new(api_secret.encode('utf-8'), signature_origin.encode
                    ('utf-8'), digestmod=hashlib.sha256).digest()
    signature_sha = base64.b64encode(signature_sha).decode(encoding='utf-8')
    authorization_origin = "api_key=\"%s\", algorithm=\"%s\", headers=\"%s\",
                            signature=\"%s\"" % (api_key, "hmac-sha256",
                            "host date request-line", signature_sha)
    authorization = base64.b64encode
(authorization_origin.encode('utf-8')).decode(encoding='utf-8')
    print(authorization_origin)
    values = {
        "host": host,
        "date": date,
        "authorization": authorization
    }
    return requset_url + "?" + urlencode(values)

def gen_body(appid, img_path, server_id):
    with open(img_path, 'rb') as f:
        img_data = f.read()
    body = {
        "header": {
            "app_id": appid,
            "status": 3
        },
        "parameter": {
            server_id: {
                "service_kind": "face_detect",
                #"detect_points": "1", #检测特征点
                #"detect_property": "1", #检测人脸属性
                "face_detect_result": {
                    "encoding": "utf8",
                    "compress": "raw",
                    "format": "json"
                }
            }
```

```
        },
        "payload": {
            "input1": {
                "encoding": "jpg",
                "status": 3,
                "image": str(base64.b64encode(img_data), 'utf-8')
            }
        }
    }
    return json.dumps(body)


def run(appid, apikey, apisecret, img_path, server_id='s67c9c78c'):
    url = 'http://api.xf-yun.com/v1/private/{}'.format(server_id)
    request_url = assemble_ws_auth_url(url, "POST", apikey, apisecret)
    headers = {'content-type': "application/json", 'host': 'api.xf-yun.com',
            'app_id': appid}
    print(request_url)
    response = requests.post(request_url, data=gen_body(appid, img_path,
                            server_id), headers=headers)
    resp_data = json.loads(response.content.decode('utf-8'))
    print(resp_data)
    print(base64.b64decode
(resp_data['payload']['face_detect_result']['text']).decode())

#填写控制台获取的APPID、APISecret、APIKey以及要检测的图片路径
if __name__ == '__main__':
    run(
        appid='11b6f6bb',
        apisecret='    ',
        apikey='     ',
        img_path=r'111.jpg',
    )
```

执行代码后会返回照片 111.jpg 的基本信息：

```
Tue, 15 Jun 2021 07:19:51 GMT
host: api.xf-yun.com
date: Tue, 15 Jun 2021 07:19:51 GMT
POST /v1/private/s67c9c78c HTTP/1.1
api_key="431ee6a26215785be1c4c75c61178a06", algorithm="hmac-sha256",
headers="host date request-line",
signature="+aQNtQj7B7mOw0Y3TFhTArsSzGvuH01woPDTTsZtUO0="
http://api.xf-yun.com/v1/private/s67c9c78c?host=api.xf-yun.com&date=Tue%2C+
15+Jun+2021+07%3A19%3A51+GMT&authorization=YXBpX2tleT0iNDMxZWU2YTI2MjE1Nzg1
YmUxYzRjNzVjNjExNzhhMDYiLCBhbGdvcml0aG09ImhtYWMtc2hhMjU2IiwgaGVhZGVycz0iaG9
zdCBkYXRlIHJlcXVlc3QtbGluZSIsIHNpZ25hdHVyZT0iK2FRTnRRajdCN21PdzBZM1RGaFRBcn
NTekd2dUgwMXdvUERUVHNadFVPMD0i
```

```
{'header': {'code': 0, 'message': 'success', 'sid':
'ase000da92c@hu17a0e8a5c330210882'}, 'payload': {'face_detect_result':
{'compress': 'raw', 'encoding': 'utf8', 'format': 'json', 'text':
'ewoJInJldCIgOiAwLAoJImZhY2VfbnVtIiA6IDEsCgkiZmFjZV8xIiA6IAoJewoJCSJ4IiA6ID
QyLAoJCSJ5IiA6IDIxMSwKCQkidyIgOiAyMDUsCgkJImgiIDogMjQ4LAoJCSJzY29yZSIgOiAwL
jk4MjM0MDQ1NTA1NTIzNjgyCgl9Cn0K'}}}
{
    "ret" : 0,
    "face_num" : 1,
    "face_1" :
    {
        "x" : 42,
        "y" : 211,
        "w" : 205,
        "h" : 248,
        "score" : 0.98234045505523682
    }
}
```

再看下面的实例文件 face_age_keda.py，功能是基于图普的深度学习算法，检测指定URL图像中的人脸并进行年龄预测。

```
# 人脸特征分析年龄 webapi 接口地址
URL = "http://tupapi.xfyun.cn/v1/age"
# 应用 ID  (必须为 webapi 类型应用，并人脸特征分析服务，参考帖子如何创建一个 webapi 应用：
# http://bbs.xfyun.cn/forum.php?mod=viewthread&tid=36481)
APPID = "11b6f6bb"
# 接口密钥(webapi 类型应用开通人脸特征分析服务后，控制台--我的应用---人脸特征分析---
# 服务的 apikey)
API_KEY = "431ee6a26215785be1c4c75c61178a06"
ImageName = "girl"
ImageUrl = "https://gimg2.baidu.com/image_search/src=http%3A%2F%2Fn.sinaimg.
cn%2Fsinacn10%2F309%2Fw534h575%2F20180926%2Fa837-hhuhisn1021919.jpg&refer=h
ttp%3A%2F%2Fn.sinaimg.cn&app=2002&size=f9999,10000&q=a80&n=0&g=0n&fmt=jpeg?
sec=1626338751&t=f087fe8fbe7dd7d66543003d8312f9dc"
# 图片数据可以通过两种方式上传，第一种在请求头设置 image_url 参数，第二种将图片二进制数据
# 写入请求体中。若同时设置，以第一种为准
# 此 demo 使用第一种方式进行上传图片地址，如果想使用第二种方式，将图片二进制数据写入请求体即可

def getHeader(image_name, image_url=None):
    curTime = str(int(time.time()))
    param = "{\"image_name\":\"" + image_name + "\",\"image_url\":\"" + image_url + "\"}"
    paramBase64 = base64.b64encode(param.encode('utf-8'))
    tmp = str(paramBase64, 'utf-8')

    m2 = hashlib.md5()
    m2.update((API_KEY + curTime + tmp).encode('utf-8'))
```

```
    checkSum = m2.hexdigest()

    header = {
        'X-CurTime': curTime,
        'X-Param': paramBase64,
        'X-Appid': APPID,
        'X-CheckSum': checkSum,
    }
    return header

r = requests.post(URL, headers=getHeader(ImageName, ImageUrl))
print(r.content)
```

执行代码后会返回解析网络图片 ImageUrl 年龄的信息，返回的是 JSON 格式的结果：


```
b'{"code":0,"data":{"fileList":[{"label":5,"labels":[5,12,4,6,11],"name":"h
ttps://gimg2.baidu.com/image_search/src=http%3A%2F%2Fn.sinaimg.cn%2Fsinacn1
0%2F309%2Fw534h575%2F20180926%2Fa837-hhuhisn1021919.jpg\\u0026refer=http%3A
%2F%2Fn.sinaimg.cn\\u0026app=2002\\u0026size=f9999,10000\\u0026q=a80\\u0026
n=0\\u0026g=0n\\u0026fmt=jpeg?sec=1626338751\\u0026t=f087fe8fbe7dd7d6654300
3d8312f9dc","rate":0.5975701808929443,"rates":[0.5975701808929443,0.2697626
9483566284,0.0613187737762928,0.03654952719807625,0.020218132063746452],"re
view":false,"tag":"Using
url"}],"reviewCount":0,"topNStatistic":[{"count":1,"label":5}]},"desc":"suc
cess","sid":"tup00002d5b@dx3d361423b5851aba00"}'
```


在上述返回的 JSON 格式的结果中，label 表示对年龄的预测，不同 label 值对应的年龄字段信息如表 8-1 所示。

表 8-1　label 对应的年龄段

label 值	对应年龄段(岁)	label 值	对应年龄段(岁)
0	0～1	7	41～50
1	2～5	8	51～60
2	6～10	9	61～80
3	11～15	10	80 以上
4	16～20	11	其他
5	21～25	12	26～30
6	31～40		

这就说明，科大讯飞预测网络图片 ImageUrl 的年龄范围是 21~25。

第9章

AI人脸识别签到打卡系统(PyQt5+百度AI+OpenCV-Python+SQLite3实现)

为了提高上课签到的效率，实现无纸化办公的需求，某高校决定参考钉钉打卡软件开发一个基于人脸识别的学生签到打卡系统。在本章的内容中，将详细讲解使用 PyQt5、百度智能云、Opencv-Python 和 SQLite3 实现一个学生签到打卡系统的过程。

9.1 需求分析

编写需求分析文档的目的是说明运行本签到系统的最终条件、性能要求及要实现的功能，为进一步设计与实现打下基础。它以文档形式将用户对软件的需求固定下来，是与用户沟通的成果，也供用户验收项目时参考。本文档预期读者为：用户，项目管理人员，软件设计人员，编程人员，测试人员等项目相关人员。

扫码观看本节视频讲解

9.1.1 背景介绍

为了保证现在大学生课堂出勤率以及学生信息的管理，大部分学校纷纷采取不同的措施来对学生的出勤率进行管理和安排，故对合理、高利用率的学生签到系统有着迫切的需求，一个可行的系统对此有着重要的意义。

- 系统用途：本系统利用相应的安卓平台，帮助学校等各个部门更加电子化、智能化地管理学生出勤的运作，从而提高学校管理的效率。
- 系统开发人员：本系统由××团队完成从可行性分析、需求分析、概要设计、实现、调试等一系列过程。

9.1.2 任务目标

(1) 用户组：为了更加清晰明了地管理，将学生信息进行分类，将每个班级的学生信息作为一个分类。

- 添加用户组：添加代表某个班级的新的用户组。
- 删除用户组：删除一个已经存在的用户组。
- 查询用户组：根据输入关键字查询某个用户组。

(2) 用户：每个用户对应一个学生信息。

- 添加用户：打开摄像头，向系统中添加新的学生信息，包括添加学生的人脸照片、学号、姓名和班级等。
- 删除用户：删除一个已经存在的用户信息。
- 更新用户：修改一个已经存在的用户信息。

(3) 签到：实现人脸识别签到功能。

- 启动签到：打开摄像头，识别当前人的信息，如果在系统数据库中存在此人的照片信息，则签到成功并显示此人的基本信息。

- 导出信息：导出此人的签到信息。
- 关闭签到：关闭当前的签到功能。

9.2 模块架构

在开发一个大型应用程序时，组织模块架构是一个非常重要的前期准备工作，是关系到整个项目的实现流程是否顺利完成的关键。在本节的内容中，将根据严格的市场需求分析，得出我们这个项目的模块结构。本系统的基本模块架构如图 9-1 所示。

扫码观看本节视频讲解

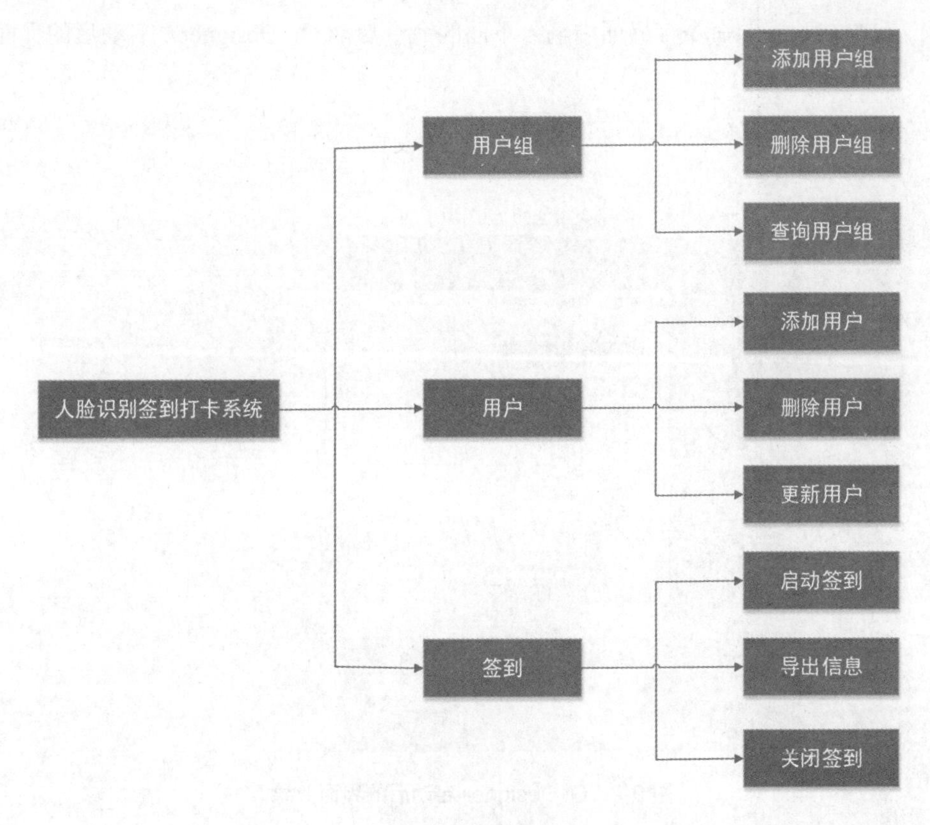

图 9-1　模块架构

9.3 使用 Qt Designer 实现主窗口界面

Qt Designer 是交互式可视化 GUI 设计工具，可以帮助我们提高开发 PyQt 程序的速度。Qt Designer 生成的 UI 界面是一个后缀为“.ui”的文件，可以通过 PyIuc 工具转换为“.py”文件。在本节的内容中，将详细讲解使用 Qt Designer 为本项目设计主窗口界面的过程。

扫码观看本节视频讲解

9.3.1 设计系统 UI 主界面

(1) 在使用 Qt Designer 设计窗口界面之前，需要先使用如下命令安装 PyQt5-tools：

```
pip install PyQt5-tools
```

(2) 在 Pycharm 的命令行界面中输入下面的命令启动 Qt Designer，启动后的界面效果如图 9-2 所示。

```
pyqt5designer.exe
```

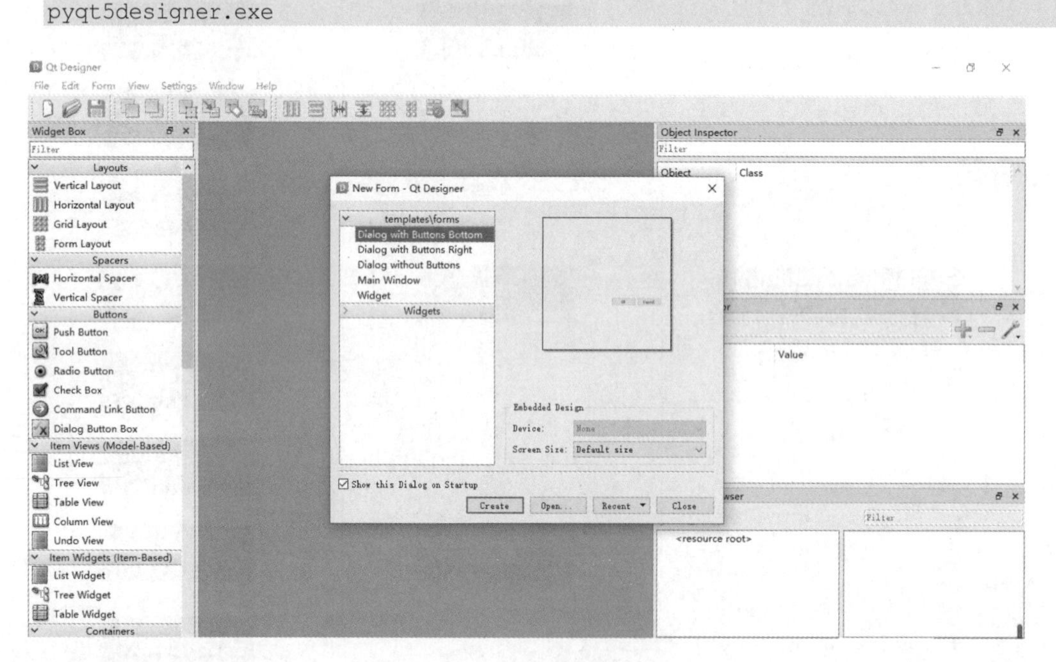

图 9-2　Qt Designer 启动后的界面效果

(3) 选择 Main Window 类型，然后单击 Create 按钮创建一个界面文件，在界面文件中设计 3 个主菜单“签到”“用户组”和“用户”，并分别为主菜单设计对应的子菜单。系统主界面文件 mainwindow.ui 的最终效果如图 9-3 所示。

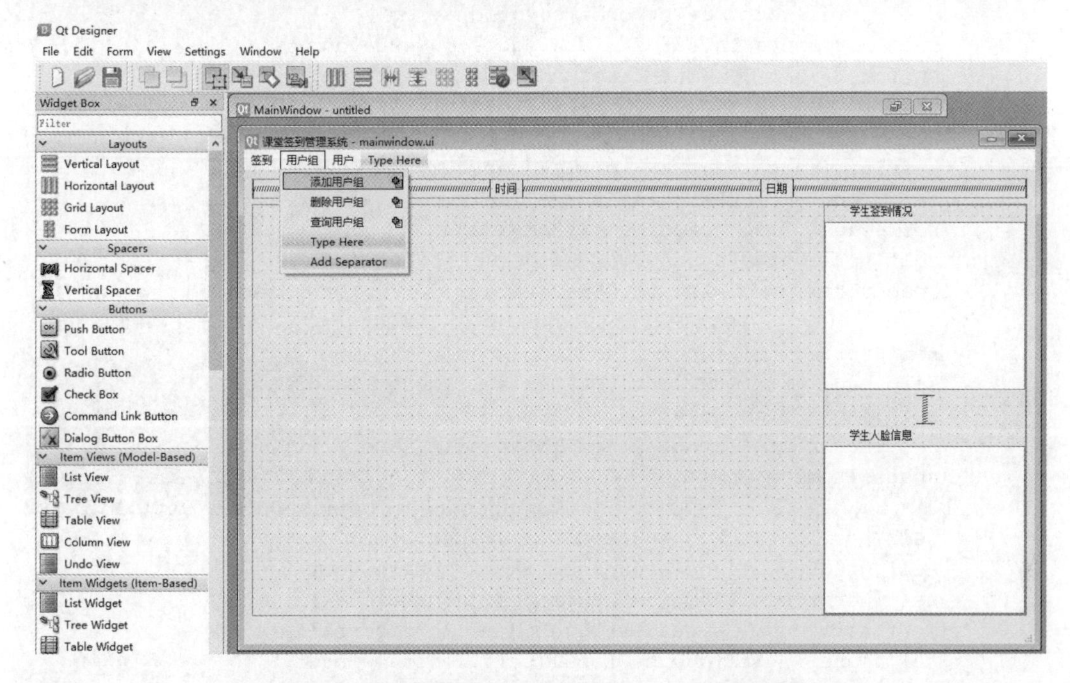

图 9-3 系统主界面文件 mainwindow.ui 的最终效果

9.3.2 将 Qt Designer 文件转换为 Python 文件

因为 Python 只会解释执行“.py”文件，不能识别“.ui”类型的文件，所以接下来需要把“.ui”类型的文件转换成“.py”文件。在 Pycharm 的命令行界面中输入下面的命令启动 Qt Designer，将 Qt Designer 设计文件 mainwindow.ui 转换成 Python 文件 mainwindow.py。

```
pyuic5 mainwindow.ui  -o  mainwindow.py
```

Python 文件 mainwindow.py 的具体实现代码如下所示：

```
from PyQt5 import QtCore, QtGui, QtWidgets

class Ui_MainWindow(object):
   def setupUi(self, MainWindow):
      MainWindow.setObjectName("MainWindow")
      MainWindow.resize(900, 545)
      self.centralwidget = QtWidgets.QWidget(MainWindow)
```

```
self.centralwidget.setMinimumSize(QtCore.QSize(900, 500))
self.centralwidget.setObjectName("centralwidget")
self.gridLayout = QtWidgets.QGridLayout(self.centralwidget)
self.gridLayout.setObjectName("gridLayout")
self.horizontalLayout_2 = QtWidgets.QHBoxLayout()
self.horizontalLayout_2.setObjectName("horizontalLayout_2")
spacerItem = QtWidgets.QSpacerItem(40, 20,
    QtWidgets.QSizePolicy.Expanding, QtWidgets.QSizePolicy.Minimum)
self.horizontalLayout_2.addItem(spacerItem)
self.label_2 = QtWidgets.QLabel(self.centralwidget)
self.label_2.setObjectName("label_2")
self.horizontalLayout_2.addWidget(self.label_2)
spacerItem1 = QtWidgets.QSpacerItem(40, 20, QtWidgets.
            QSizePolicy.Expanding, QtWidgets.QSizePolicy.Minimum)
self.horizontalLayout_2.addItem(spacerItem1)
self.label_3 = QtWidgets.QLabel(self.centralwidget)
self.label_3.setObjectName("label_3")
self.horizontalLayout_2.addWidget(self.label_3)
spacerItem2 = QtWidgets.QSpacerItem(40, 20, QtWidgets.QSizePolicy.
                        Expanding, QtWidgets.QSizePolicy.Minimum)
self.horizontalLayout_2.addItem(spacerItem2)
self.gridLayout.addLayout(self.horizontalLayout_2, 0, 0, 1, 1)
self.horizontalLayout = QtWidgets.QHBoxLayout()
self.horizontalLayout.setObjectName("horizontalLayout")
self.label = QtWidgets.QLabel(self.centralwidget)
self.label.setMinimumSize(QtCore.QSize(640, 480))
self.label.setText("")
self.label.setObjectName("label")
self.horizontalLayout.addWidget(self.label)
self.verticalLayout_2 = QtWidgets.QVBoxLayout()
self.verticalLayout_2.setObjectName("verticalLayout_2")
self.label_4 = QtWidgets.QLabel(self.centralwidget)
self.label_4.setObjectName("label_4")
self.verticalLayout_2.addWidget(self.label_4)
self.plainTextEdit = QtWidgets.QPlainTextEdit(self.centralwidget)
self.plainTextEdit.setMinimumSize(QtCore.QSize(200, 0))
self.plainTextEdit.setLineWidth(1)
self.plainTextEdit.setReadOnly(True)
self.plainTextEdit.setTabStopWidth(80)
self.plainTextEdit.setObjectName("plainTextEdit")
self.verticalLayout_2.addWidget(self.plainTextEdit)
spacerItem3 = QtWidgets.QSpacerItem(20, 40, QtWidgets.QSizePolicy.
                        Minimum, QtWidgets.QSizePolicy.Expanding)
self.verticalLayout_2.addItem(spacerItem3)
self.label_5 = QtWidgets.QLabel(self.centralwidget)
```

```
self.label_5.setObjectName("label_5")
self.verticalLayout_2.addWidget(self.label_5)
self.plainTextEdit_2 = QtWidgets.QPlainTextEdit(self.centralwidget)
self.plainTextEdit_2.setMinimumSize(QtCore.QSize(200, 0))
self.plainTextEdit_2.setObjectName("plainTextEdit_2")
self.verticalLayout_2.addWidget(self.plainTextEdit_2)
self.horizontalLayout.addLayout(self.verticalLayout_2)
self.horizontalLayout.setStretch(1, 1)
self.gridLayout.addLayout(self.horizontalLayout, 1, 0, 1, 1)
MainWindow.setCentralWidget(self.centralwidget)
self.menubar = QtWidgets.QMenuBar(MainWindow)
self.menubar.setGeometry(QtCore.QRect(0, 0, 900, 23))
self.menubar.setObjectName("menubar")
self.menu = QtWidgets.QMenu(self.menubar)
self.menu.setObjectName("menu")
self.menu_2 = QtWidgets.QMenu(self.menubar)
self.menu_2.setObjectName("menu_2")
self.menu_3 = QtWidgets.QMenu(self.menubar)
self.menu_3.setObjectName("menu_3")
MainWindow.setMenuBar(self.menubar)
self.statusbar = QtWidgets.QStatusBar(MainWindow)
self.statusbar.setObjectName("statusbar")
MainWindow.setStatusBar(self.statusbar)
self.actionopen = QtWidgets.QAction(MainWindow)
self.actionopen.setObjectName("actionopen")
self.actionclose = QtWidgets.QAction(MainWindow)
self.actionclose.setObjectName("actionclose")
self.actionaddgroup = QtWidgets.QAction(MainWindow)
self.actionaddgroup.setObjectName("actionaddgroup")
self.actiondelgroup = QtWidgets.QAction(MainWindow)
self.actiondelgroup.setObjectName("actiondelgroup")
self.actionsave = QtWidgets.QAction(MainWindow)
self.actionsave.setObjectName("actionsave")
self.actiond = QtWidgets.QAction(MainWindow)
self.actiond.setObjectName("actiond")
self.actiongetlist = QtWidgets.QAction(MainWindow)
self.actiongetlist.setObjectName("actiongetlist")
self.actionadduser1 = QtWidgets.QAction(MainWindow)
self.actionadduser1.setObjectName("actionadduser1")
self.actionadduser = QtWidgets.QAction(MainWindow)
self.actionadduser.setObjectName("actionadduser")
self.actiondeluser = QtWidgets.QAction(MainWindow)
self.actiondeluser.setObjectName("actiondeluser")
self.actionupdateuser = QtWidgets.QAction(MainWindow)
self.actionupdateuser.setObjectName("actionupdateuser")
```

```
        self.menu.addAction(self.actionopen)
        self.menu.addAction(self.actionclose)
        self.menu_2.addAction(self.actionaddgroup)
        self.menu_2.addAction(self.actiondelgroup)
        self.menu_2.addAction(self.actiongetlist)
        self.menu_3.addAction(self.actionadduser)
        self.menu_3.addAction(self.actiondeluser)
        self.menu_3.addAction(self.actionupdateuser)
        self.menubar.addAction(self.menu.menuAction())
        self.menubar.addAction(self.menu_2.menuAction())
        self.menubar.addAction(self.menu_3.menuAction())

        self.retranslateUi(MainWindow)
        QtCore.QMetaObject.connectSlotsByName(MainWindow)

    def retranslateUi(self, MainWindow):
        _translate = QtCore.QCoreApplication.translate
        MainWindow.setWindowTitle(_translate("MainWindow", "课堂签到管理系统"))
        self.label_2.setText(_translate("MainWindow", "时间"))
        self.label_3.setText(_translate("MainWindow", "日期"))
        self.label_4.setText(_translate("MainWindow", "    学生签到情况"))
        self.label_5.setText(_translate("MainWindow", "    学生人脸信息"))
        self.menu.setTitle(_translate("MainWindow", "签到"))
        self.menu_2.setTitle(_translate("MainWindow", "用户组"))
        self.menu_3.setTitle(_translate("MainWindow", "用户"))
        self.actionopen.setText(_translate("MainWindow", "启动签到"))
        self.actionclose.setText(_translate("MainWindow", "关闭签到"))
        self.actionaddgroup.setText(_translate("MainWindow", "添加用户组"))
        self.actiondelgroup.setText(_translate("MainWindow", "删除用户组"))
        self.actionsave.setText(_translate("MainWindow", "存储信息"))
        self.actiond.setText(_translate("MainWindow", "导出信息"))
        self.actiongetlist.setText(_translate("MainWindow", "查询用户组"))
        self.actionadduser1.setText(_translate("MainWindow", "添加用户"))
        self.actionadduser.setText(_translate("MainWindow", "添加用户"))
        self.actiondeluser.setText(_translate("MainWindow", "删除用户"))
        self.actionupdateuser.setText(_translate("MainWindow", "更新用户"))
```

编写文件 main.py，功能是调用上面的 UI 界面文件创建应用程序对象，然后分别创建并显示系统主界面的窗口，代码如下：

```
import sys
from PyQt5.QtWidgets import QApplication

from mywindow import mywindow
```

```
if __name__ == '__main__':
    # 创建应用程序对象
    app = QApplication(sys.argv)
    # 创建窗口
    ui = mywindow()
    # 显示窗口
    ui.show()
    # 应用执行
    app.exec_()
    sys.exit(0)
```

运行程序文件 main.py 会显示系统主界面，最终执行效果如图 9-4 所示。

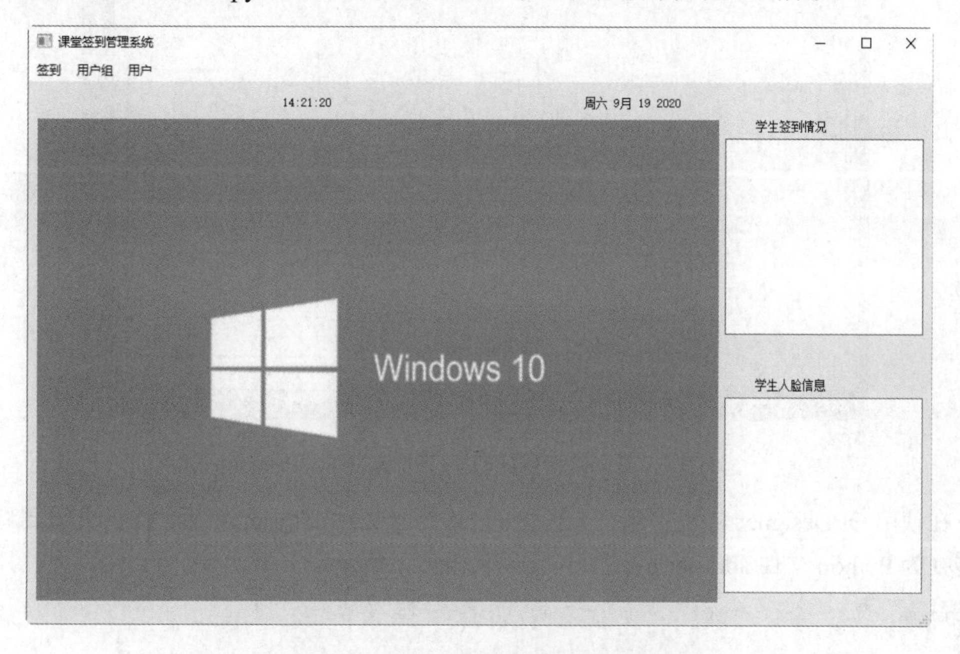

图 9-4　系统主界面的最终执行效果

9.4　签到打卡、用户操作和用户组操作

签到打卡、用户操作和用户组操作是本项目的核心，首先使用 Qt Designer 分别设计一个打卡界面、用户操作界面和用户组操作界面，再实现一个摄像头类打开摄像头，然后实现实时显示时间功能，并调用百度 AI 模块实现人脸识别功能。

扫码观看本节视频讲解

9.4.1 设计 UI 界面

使用 Qt Designer 设计签到打卡模块的 UI 界面文件 adduser.ui，在界面左侧显示摄像头打卡界面，在右侧分别显示签到情况和识别的人脸信息，如图 9-5 所示。

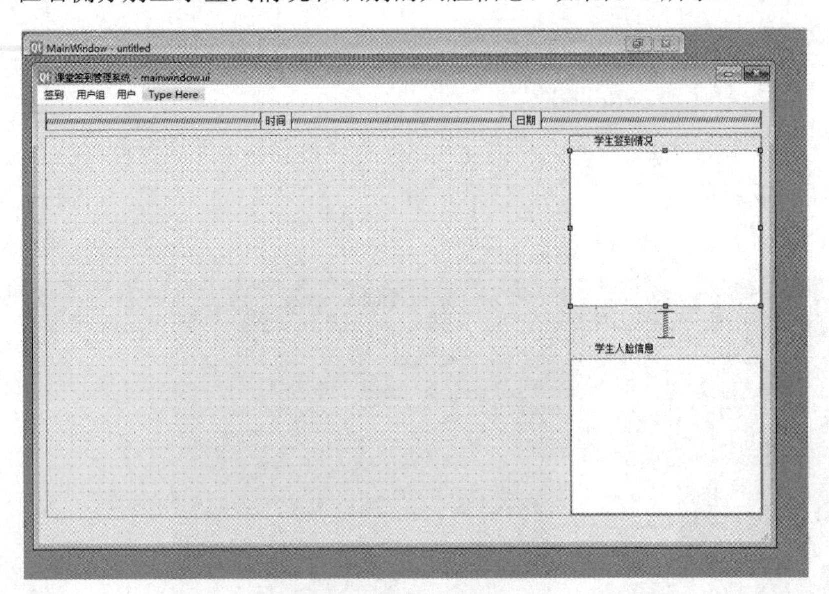

图 9-5　签到打卡模块的 UI 界面文件 adduser.ui

在使用 Qt Designer 实现签到打卡模块的 UI 界面文件 adduser.ui 后，接下来需要将此文件转换为 Python 文件 adduser.py，具体实现代码如下所示：

```
from PyQt5 import QtCore, QtGui, QtWidgets

class Ui_Dialog(object):
   def setupUi(self, Dialog):
      Dialog.setObjectName("Dialog")
      Dialog.resize(445, 370)
      self.gridLayout = QtWidgets.QGridLayout(Dialog)
      self.gridLayout.setObjectName("gridLayout")
      self.verticalLayout_4 = QtWidgets.QVBoxLayout()
      self.verticalLayout_4.setObjectName("verticalLayout_4")
      self.horizontalLayout = QtWidgets.QHBoxLayout()
      self.horizontalLayout.setObjectName("horizontalLayout")
      self.verticalLayout = QtWidgets.QVBoxLayout()
      self.verticalLayout.setObjectName("verticalLayout")
      self.label = QtWidgets.QLabel(Dialog)
```

```
self.label.setMinimumSize(QtCore.QSize(305, 225))
self.label.setText("")
self.label.setObjectName("label")
self.verticalLayout.addWidget(self.label)
self.pushButton = QtWidgets.QPushButton(Dialog)
self.pushButton.setObjectName("pushButton")
self.verticalLayout.addWidget(self.pushButton)
self.horizontalLayout.addLayout(self.verticalLayout)
self.groupBox = QtWidgets.QGroupBox(Dialog)
self.groupBox.setObjectName("groupBox")
self.listWidget = QtWidgets.QListWidget(self.groupBox)
self.listWidget.setGeometry(QtCore.QRect(10, 20, 81, 231))
self.listWidget.setObjectName("listWidget")
self.horizontalLayout.addWidget(self.groupBox)
self.horizontalLayout.setStretch(0, 3)
self.horizontalLayout.setStretch(1, 1)
self.verticalLayout_4.addLayout(self.horizontalLayout)
self.verticalLayout_3 = QtWidgets.QVBoxLayout()
self.verticalLayout_3.setObjectName("verticalLayout_3")
self.verticalLayout_2 = QtWidgets.QVBoxLayout()
self.verticalLayout_2.setObjectName("verticalLayout_2")
self.horizontalLayout_2 = QtWidgets.QHBoxLayout()
self.horizontalLayout_2.setObjectName("horizontalLayout_2")
self.label_2 = QtWidgets.QLabel(Dialog)
self.label_2.setObjectName("label_2")
self.horizontalLayout_2.addWidget(self.label_2)
self.lineEdit = QtWidgets.QLineEdit(Dialog)
self.lineEdit.setObjectName("lineEdit")
self.horizontalLayout_2.addWidget(self.lineEdit)
self.verticalLayout_2.addLayout(self.horizontalLayout_2)
self.horizontalLayout_3 = QtWidgets.QHBoxLayout()
self.horizontalLayout_3.setObjectName("horizontalLayout_3")
self.label_3 = QtWidgets.QLabel(Dialog)
self.label_3.setObjectName("label_3")
self.horizontalLayout_3.addWidget(self.label_3)
self.lineEdit_2 = QtWidgets.QLineEdit(Dialog)
self.lineEdit_2.setObjectName("lineEdit_2")
self.horizontalLayout_3.addWidget(self.lineEdit_2)
self.label_4 = QtWidgets.QLabel(Dialog)
self.label_4.setObjectName("label_4")
self.horizontalLayout_3.addWidget(self.label_4)
self.lineEdit_3 = QtWidgets.QLineEdit(Dialog)
self.lineEdit_3.setObjectName("lineEdit_3")
self.horizontalLayout_3.addWidget(self.lineEdit_3)
self.verticalLayout_2.addLayout(self.horizontalLayout_3)
self.verticalLayout_3.addLayout(self.verticalLayout_2)
self.horizontalLayout_4 = QtWidgets.QHBoxLayout()
```

```
    self.horizontalLayout_4.setObjectName("horizontalLayout_4")
    spacerItem = QtWidgets.QSpacerItem(40, 20, QtWidgets.QSizePolicy.
                Expanding, QtWidgets.QSizePolicy.Minimum)
    self.horizontalLayout_4.addItem(spacerItem)
    self.pushButton_2 = QtWidgets.QPushButton(Dialog)
    self.pushButton_2.setObjectName("pushButton_2")
    self.horizontalLayout_4.addWidget(self.pushButton_2)
    spacerItem1 = QtWidgets.QSpacerItem(40, 20, QtWidgets.QSizePolicy.
                Expanding, QtWidgets.QSizePolicy.Minimum)
    self.horizontalLayout_4.addItem(spacerItem1)
    self.pushButton_3 = QtWidgets.QPushButton(Dialog)
    self.pushButton_3.setObjectName("pushButton_3")
    self.horizontalLayout_4.addWidget(self.pushButton_3)
    spacerItem2 = QtWidgets.QSpacerItem(40, 20, QtWidgets.QSizePolicy.
                Expanding, QtWidgets.QSizePolicy.Minimum)
    self.horizontalLayout_4.addItem(spacerItem2)
    self.verticalLayout_3.addLayout(self.horizontalLayout_4)
    self.verticalLayout_4.addLayout(self.verticalLayout_3)
    self.verticalLayout_4.setStretch(0, 3)
    self.verticalLayout_4.setStretch(1, 1)
    self.gridLayout.addLayout(self.verticalLayout_4, 0, 0, 1, 1)

    self.retranslateUi(Dialog)
    QtCore.QMetaObject.connectSlotsByName(Dialog)

  def retranslateUi(self, Dialog):
    _translate = QtCore.QCoreApplication.translate
    Dialog.setWindowTitle(_translate("Dialog", "添加用户"))
    self.pushButton.setText(_translate("Dialog", "选择人脸图片"))
    self.groupBox.setTitle(_translate("Dialog", "用户组选择"))
    self.label_2.setText(_translate("Dialog", "输入学号(字母数组下画线): "))
    self.label_3.setText(_translate("Dialog", "姓名: "))
    self.label_4.setText(_translate("Dialog", "班级: "))
    self.pushButton_2.setText(_translate("Dialog", "确定"))
    self.pushButton_3.setText(_translate("Dialog", "取消"))
```

9.4.2 创建摄像头类

编写文件 cameravideo.py，创建一个摄像头类 camera，分别实现如下功能：

- 打开摄像头。
- 获取摄像头的实时数据。
- 将摄像头的数据进行转换并提供给界面。

文件 cameravideo.py 的具体实现代码如下所示：

```
import cv2
import numpy as np
from  PyQt5.QtGui import QPixmap,QImage

class camera():
    def __init__(self):
        #类 VideoCapture：调用摄像头进行读取视频内容
        #下面的参数 0 表示默认打开摄像头
        #self.capture 表示打开的摄像头对象
        self.capture = cv2.VideoCapture(0, cv2.CAP_DSHOW)
        #isOpened 函数返回一个布尔值，来判断是否摄像头初始化成功
        # if self.capture.isOpened():
        #     print("isOpened")
        #定义一个多维数组，存取画面
        self.currentframe = np.array([])

    #读取摄像头数据
    def read_camera(self):
        #ret 是否成功，pic_data 数据
        ret,data = self.capture.read()
        if not ret:
            print("获取摄像头数据失败")
            return None
        return data

    #数据转换成界面能显示的数据格式
    def camera_to_pic(self):
        pic = self.read_camera()
        #摄像头是 BGR 方式存储，首先要转换成 RGB
        self.currentframe = cv2.cvtColor(pic,cv2.COLOR_BGR2RGB)
        #设置宽高
        #self.currentframe = cv2.cvtColor(self.currentframe,(640,480))

        #转换格式(界面能够显示的格式)
        #获取画面的宽度和高度
        height,width = self.currentframe.shape[:2]
        #先转换成 QImage 类型的图片(画面)，创建 QImage 类对象，使用摄像头的画面
        #QImage (data, width, height , format)创建:数据，宽度，高度,格式
        qimg = QImage(self.currentframe,width,height,QImage.Format_RGB888)
        qpixmap = QPixmap.fromImage(qimg)
        return qpixmap

    def close_camera(self):
        self.capture.release()
```

9.4.3 UI 界面的操作处理

在签到打卡模块的 UI 界面文件 adduser.py 中有许多元素，例如顶部菜单、顶部时间显示、素材背景图片和右侧的签到列表信息等。接下来编写文件 mywindow.py，为 UI 界面文件 adduser.py 中的元素实现操作处理。

(1) 创建界面类 mywindow。在 PyQt5 中存在一个机制：信号可以关联上另一个函数，另一个函数在产生这个对应信号的时候就会调用“信号槽机制”。在组件中存在一些特定的信号，当组件执行某个操作的时候，对应的信号就会被激活(信号产生)。由于信号(如点击)可能执行的功能是不一样的，我们需要指定信号与槽函数的关联，设置信号要去执行的功能是什么信号对象。在类 mywindow 的初始化代码中，首先创建一个时间定时器，并设置在创建窗口的伊始就调用方法 get_accesstoken()申请访问百度 AI 的令牌，然后设置了不同信号槽需要执行的对应功能函数。代码如下：

```
class mywindow(Ui_MainWindow,QMainWindow):
    detect_data_signal = pyqtSignal(bytes)
    group_id = pyqtSignal(str)
    camera_status = False

    def __init__(self):
        super(mywindow,self).__init__()
        self.setupUi(self)
        self.label.setScaledContents(True)
        self.label.setPixmap(QPixmap("1.jpg"))
        # 创建一个时间定时器
        self.datetime = QTimer()
        #启动获取系统时间/日期定时器,定时时间为 10ms，每 10ms 产生一次信号
        self.datetime.start(10)
        #创建窗口就应该完成进行访问令牌的申请(获取)
        self.get_accesstoken()
        #信号与槽的关联
        #self.actionopen：指定对象
        #triggered：信号
        #connect：关联(槽函数)
        #self.on_actionopen():关联的函数
        self.actionopen.triggered.connect(self.on_actionopen)
        self.actionclose.triggered.connect(self.on_actionclose)


        #添加用户组信号槽
        self.actionaddgroup.triggered.connect(self.add_group)
```

```
        #删除用户组信号槽
        self.actiondelgroup.triggered.connect(self.del_group)
        #查询用户组信号槽
        self.actiongetlist.triggered.connect(self.getgrouplist)
        #添加用户信号槽
        self.actionadduser.triggered.connect(self.add_user)
        #删除用户信号槽
        self.actiondeluser.triggered.connect(self.del_user)
        # 更新用户信号槽
        self.actionupdateuser.triggered.connect(self.update_user)
        #关联时间/日期的定时器信号与槽函数
        self.datetime.timeout.connect(self.data_time)
```

(2) 创建定时器函数 data_time(self)，获取当前的日期与时间，并添加到对应的定时器中。在前面的类 mywindow 的初始化代码中，设置本项目的定时时间为 10ms，即每 10ms 产生一次信号。代码如下：

```
def data_time(self):
    # 获取日期
    date = QDate.currentDate()
    #print(date)
    #self.dateEdit.setDate(date)
    self.label_3.setText(date.toString())
    # 获取时间
    time = QTime.currentTime()
    #print(time)
    #self.timeEdit.setTime(time)
    self.label_2.setText(time.toString())
    # 获取日期时间
    datetime = QDateTime.currentDateTime()
    #print(datetime)
```

(3) 创建启动打卡签到函数 on_actionopen(self)，首先弹出一个对话框让用户输入所在的用户组，并不断调用摄像头识别当前打卡人员的脸部信息。代码如下：

```
def on_actionopen(self):
    list = self.getlist()
    self.group_id, ret = QInputDialog.getText(self, "请输入所在用户组",
                    "用户组信息\n" + str(list['result']['group_id_list']))
    if self.group_id == '':
        QMessageBox.about(self, "签到失败", "用户组不能为空\n")
        return
    group_status = 0
    for i in list['result']['group_id_list']:
        if i == self.group_id:
            group_status =1
```

```
            break
        if group_status == 0:
            QMessageBox.about(self, "签到失败", "该用户组不存在\n")
            return
        #启动摄像头
        self.cameravideo = camera()
        self.camera_status = True
        #启动定时器，进行定时，每个多长时间进行一次获取摄像头数据进行显示，用作流畅显示画面
        self.timeshow = QTimer(self)
        self.timeshow.start(10)
        #10ms 的定时器启动，每到 10ms 就会产生一个信号 timeout,信号没有()
        self.timeshow.timeout.connect(self.show_cameradata)
        #self.timeshow.timeout().connect(self.show_cameradata)
        # self.show_cameradata()
        #创建检测线程
        self.create_thread()
        # self.group_id.emit(str(group_id))
        # self.group_id.connect(self.detectThread.get_group_id)
        #当开启启动签到时，创建定时器，500ms，用作获取要检测的画面
        # facedetecttime 定时器设置检测画面获取
        self.facedetecttime = QTimer(self)
        self.facedetecttime.start(500)
        self.facedetecttime.timeout.connect(self.get_cameradata)

        self.detect_data_signal.connect(self.detectThread.get_base64)
        self.detectThread.transmit_data.connect(self.get_detectdata)
        self.detectThread.search_data.connect(self.get_search_data)
```

(4) 创建关闭打卡签到函数 on_actionclose(self)，功能是关闭定时器和摄像头，并显示本次签到情况是否成功。代码如下：

```
    def on_actionclose(self):
        # 清除学生人脸信息(False)
        # self.plainTextEdit_2.setPlainText("  ")
        #关闭定时器，不再设置检测画面获取
        self.facedetecttime.stop()
        #self.facedetecttime.timeout.disconnect(self.get_cameradata)
        #self.detect_data_signal.disconnect(self.detectThread.get_base64)
        #self.detectThread.transmit_data.connect(self.get_detectdata)
        #关闭检测线程
        self.detectThread.OK = False
        self.detectThread.quit()
        self.detectThread.wait()
        print(self.detectThread.isRunning())
        # 关闭定时器，不再去获取摄像头的数据
        self.timeshow.stop()
```

```
        self.timeshow.timeout.disconnect(self.show_cameradata)
        # 关闭摄像头
        self.cameravideo.close_camera()
        self.camera_status = False
        print("1")
        #显示本次签到情况
        self.signdata = sign_data(self.detectThread.sign_list,self)
        self.signdata.exec_()
        if self.timeshow.isActive() == False and self.facedetecttime.isActive()
                                     == False:
            # 画面设置为初始状态
            self.label.setPixmap(QPixmap("1.jpg"))
            self.plainTextEdit.clear()
            self.plainTextEdit_2.clear()
        else:
            QMessageBox.about(self, "错误", "关闭签到失败\n")
```

(5) 创建函数 create_thread()，功能是调用函数 detect_thread()使用多线程技术实现人脸识别功能，函数 detect_thread()在文件 detect.py 中定义实现。函数 create_thread()的代码如下：

```
    #创建线程完成检测
    def create_thread(self):
        self.detectThread = detect_thread(self.access_token,self.group_id)
        self.detectThread.start()
```

(6) 编写函数 get_cameradata(self)获取摄像头中的画面信息，然后把摄像头画面转换成图片，再将图片编码为 base64 格式的数据进行信号传递。代码如下：

```
    def get_cameradata(self):
        # 摄像头获取画面
        camera_data = self.cameravideo.read_camera()
        # 把摄像头画面转换成图片，然后设置 base64 编码格式数据
        _, enc = cv2.imencode('.jpg', camera_data)
        base64_image = base64.b64encode(enc.tobytes())
        #产生信号，传递数据
        self.detect_data_signal.emit(bytes(base64_image))
```

(7) 编写函数 show_cameradata()，功能是获取摄像头数据并显示出画面。代码如下：

```
    def show_cameradata(self):
        #获取摄像头数据，转换数据
        pic = self.cameravideo.camera_to_pic()
        #显示数据，显示画面
        self.label.setPixmap(pic)
```

(8) 编写函数 get_accesstoken(self)，功能是获取访问百度 AI 网络请求的访问令牌，代

码如下：

```
def get_accesstoken(self):
    #host 对象是字符串对象存储是授权的服务地址-----获取 accesstoken 的地址
    host = 'https://aip.baidubce.com/oauth/2.0/token?grant_type=
            client_credentials&client_id=OxjGDouMvcrtNa3SbHB2C146&client_
            secret=xxxxxxxxxx'
    #使用 get 函数发送网络请求，参数为网络请求的地址，执行时会产生返回请求的结果
    response = requests.get(host)
    if response:
        # print(response.json())
        data = response.json()
        self.access_token = data.get('access_token')
```

(9) 编写槽函数 get_detectdata()，功能是获取检测数据，调用百度 AI 实现检测并返回学生信息和签到信息，然后将识别结果打印到主窗口的右侧界面中。代码如下：

```
def get_detectdata(self,data):
    if data['error_code'] != 0:
        self.plainTextEdit_2.setPlainText(data['error_msg'])
        return
    elif data['error_msg'] == 'SUCCESS':
        # 在 data 字典中，键'result'对应的值才是返回的检查结果
        # data['result']才是检测结果
        # 人脸数目
        self.plainTextEdit_2.clear()
        face_num = data['result']['face_num']
        if face_num == 0:
            self.plainTextEdit_2.appendPlainText("未测到检人脸")
            return
        else:
            self.plainTextEdit_2.appendPlainText("测到检人脸")
        # 人脸信息 data['result']['face_list']，是列表，每个数据就是一个人脸信息，
        # 需要取出每个列表数据
        # 每个人脸信息：data['result']['face_list'][0~i]人脸信息字典
        for i in range(face_num):
            # 通过 for 循环，分别取出列表的每一个数据
            # data['result']['face_list'][i]，就是一个人脸信息的字典

            age = data['result']['face_list'][i]['age']
            beauty = data['result']['face_list'][i]['beauty']
            gender = data['result']['face_list'][i]['gender']['type']
            expression = data['result']['face_list'][i]['expression']['type']
            face_shape = data['result']['face_list'][i]['face_shape']['type']
            glasses = data['result']['face_list'][i]['glasses']['type']
            emotion = data['result']['face_list'][i]['emotion']['type']
```

```
        mask = data['result']['face_list'][i]['mask']['type']
        # 往窗口中添加文本，参数就是需要的文本信息
        self.plainTextEdit_2.appendPlainText("-----------------")
        self.plainTextEdit_2.appendPlainText("第" + str(i + 1)
                                             + "个学生信息：")
        self.plainTextEdit_2.appendPlainText("-----------------")
        self.plainTextEdit_2.appendPlainText("年龄：" + str(age))
        self.plainTextEdit_2.appendPlainText("颜值分数：" + str(beauty))
        self.plainTextEdit_2.appendPlainText("性别：" + str(gender))
        self.plainTextEdit_2.appendPlainText("表情：" + str(expression))
        self.plainTextEdit_2.appendPlainText("脸型：" + str(face_shape))
        self.plainTextEdit_2.appendPlainText("是否佩戴眼镜："
                                             + str(glasses))
        self.plainTextEdit_2.appendPlainText("情绪：" + str(emotion))
        if mask == 0:
            mask = "否"
        else:
            mask = "是"
        self.plainTextEdit_2.appendPlainText("是否佩戴口罩：" + str(mask))
        self.plainTextEdit_2.appendPlainText("-----------------")
```

(10) 编写函数 add_group(self)，功能是向系统中添加新的用户组信息，添加的信息会同步到百度 AI 的应用程序中。代码如下：

```
def add_group(self):
    #打开对话框，进行输入用户组
    group,ret = QInputDialog.getText(self,"添加用户组","请输入用户组(由数字、
                字母、下画线组成)")

    request_url = "https://aip.baidubce.com/rest/2.0/face/v3/faceset/group/add"
    params = {
        "group_id":group
    }
    access_token = self.access_token
    request_url = request_url + "?access_token=" + access_token
    headers = {'content-type': 'application/json'}
    response = requests.post(request_url, data=params, headers=headers)
    if response:
        message = response.json()
        if message['error_msg'] == 'SUCCESS':
            QMessageBox.about(self,"用户组创建结果","用户组创建成功")
        else:
            QMessageBox.about(self,"用户组创建结果","用户组创建失败\n"
                              +message['error_msg'])
```

(11) 编写函数 del_group(self)，功能是删除系统中已经存在的某用户组信息，会同步百

度 AI 的应用程序中删除这个用户组。代码如下：

```
def del_group(self):
    request_url = "https://aip.baidubce.com/rest/2.0/face/v3/faceset/
                  group/delete"
    list = self.getlist()
    group, ret = QInputDialog.getText(self, "用户组列表", "用户组信息\n"
                 + str(list['result']['group_id_list']))
    # 删除，需要知道哪些组
    params = {
        "group_id": group  # 要删除的用户组的 id
    }
    access_token = self.access_token
    request_url = request_url + "?access_token=" + access_token
    headers = {'content-type': 'application/json'}
    response = requests.post(request_url, data=params, headers=headers)
    if response:
        data = response.json()
        if data['error_msg'] == 'SUCCESS':
            QMessageBox.about(self, "用户组删除结果", "用户组删除成功")
        else:
            QMessageBox.about(self, "用户组删除结果", "用户组删除失败\n"
                              + data['error_msg'])
```

(12) 编写函数 getlist(self)，功能是获取系统中已经存在的用户组信息，并将这些用户组信息列表显示出来。代码如下：

```
#获取用户组
def getlist(self):
    request_url = "https://aip.baidubce.com/rest/2.0/face/v3/faceset/
                  group/getlist"

    params = {
        "start":0,"length":100
    }
    access_token = self.access_token
    request_url = request_url + "?access_token=" + access_token
    headers = {'content-type': 'application/json'}
    response = requests.post(request_url, data=params, headers=headers)
    if response:
        data = response.json()
        if data['error_msg'] == 'SUCCESS':
            return data
        else:
            QMessageBox.about(self, "获取用户组结果", "获取用户组失败\n"
                              + data['error_msg'])
```

(13) 编写函数 add_user(self)，功能是向系统中添加新的用户信息，在添加之前需要确保已经打开了摄像头，添加的信息会同步到百度 AI 的应用程序中。代码如下：

```
def add_user(self):
    request_url = "https://aip.baidubce.com/rest/2.0/face/v3/faceset/user/add"
    if self.camera_status:
        QMessageBox.about(self,"摄像头状态","摄像头已打开，正在进行人脸签到\
                        n请关闭签到，再添加用户")
        return
    list = self.getlist()
    #创建一个窗口来选择这些内容
    window = adduserwindow(list['result']['group_id_list'],self)
    #新创建窗口，通过exec()函数一直执行，阻塞执行，窗口不关闭
    #exec()函数不会退出，关闭窗口才会结束
    window_status = window.exec_()
    #判断是否点击确定进行关闭
    if window_status != 1:
        return
    #请求参数，需要获取人脸：转换人脸编码，添加组id，用户id，新用户id信息
    params = {
        "image":window.base64_image,#人脸图片
        "image_type":"BASE64",#人脸图片编码
        "group_id":window.group_id,#组id
        "user_id":window.user_id,#新用户id
        "user_info":'姓名：'+window.msg_name+'\n'
                 +'班级：'+window.msg_class#用户信息
    }
    access_token = self.access_token
    request_url = request_url + "?access_token=" + access_token
    headers = {'content-type': 'application/json'}
    response = requests.post(request_url, data=params, headers=headers)
    if response:
        data = response.json()
        if data['error_msg'] == 'SUCCESS':
            QMessageBox.about(self, "人脸创建结果", "人脸创建成功")
        else:
            QMessageBox.about(self, "人脸创建结果", "用户组创建失败\n"
                              + data['error_msg'])
```

(14) 编写函数 update_user(self)，功能是修改系统中已经存在的某用户信息，会同步在百度 AI 的应用程序中更新这个用户的信息。代码如下：

```
#更新用户人脸
def update_user(self):
    request_url = "https://aip.baidubce.com/rest/2.0/face/v3/faceset/
                user/update"
```

```
    if self.camera_status:
        QMessageBox.about(self,"摄像头状态","摄像头已打开，正在进行人脸签到\n
                                    请关闭签到，再添加用户")
        return
    list = self.getlist()
    #创建一个窗口来选择这些内容
    window = adduserwindow(list['result']['group_id_list'],self)
    #新创建窗口，通过 exec()函数一直执行，阻塞执行，窗口不进行关闭
    #exec()函数不会退出，关闭窗口才会结束
    window_status = window.exec_()
    #判断是否点击确定进行关闭
    if window_status != 1:
        return
    #请求参数，需要获取人脸：转换人脸编码，添加组 id，用户 id，新用户 id 信息
    params = {
        "image":window.base64_image,#人脸图片
        "image_type":"BASE64",#人脸图片编码
        "group_id":window.group_id,#组 id
        "user_id":window.user_id,#新用户 id
        "user_info":'姓名：'+window.msg_name+'\n'
                  +'班级：'+window.msg_class#用户信息
    }
    access_token = self.access_token
    request_url = request_url + "?access_token=" + access_token
    headers = {'content-type': 'application/json'}
    response = requests.post(request_url, data=params, headers=headers)
    if response:
        data = response.json()
        if data['error_msg'] == 'SUCCESS':
            QMessageBox.about(self, "人脸更新结果", "人脸更新成功")
        else:
            QMessageBox.about(self, "人脸更新结果", "用户组更新失败\n"
                                + data['error_msg'])
```

(15) 编写函数 get_userlist()获取系统中已经存在的用户列表信息，代码如下：

```
def get_userlist(self,group):
    request_url = "https://aip.baidubce.com/rest/2.0/face/v3/faceset/
                group/getusers"

    params = {
        "group_id":group
    }
    access_token = self.access_token
    request_url = request_url + "?access_token=" + access_token
    headers = {'content-type': 'application/json'}
    response = requests.post(request_url, data=params, headers=headers)
```

```
        if response:
            data = response.json()
            if data['error_msg'] == 'SUCCESS':
                return data
            else:
                QMessageBox.about(self, "获取用户列表结果", "获取用户列表失败\n"
                                  + data['error_msg'])
```

(16) 编写函数 user_face_list()，功能是获取系统中已经存在的用户的人脸信息，代码如下：

```
    #获取用户人脸列表
    def user_face_list(self,group,user):
        request_url = "https://aip.baidubce.com/rest/2.0/face/v3/faceset/
                       face/getlist"

        params = {
            "user_id":user,
            "group_id":group
        }
        access_token = self.access_token
        request_url = request_url + "?access_token=" + access_token
        headers = {'content-type': 'application/json'}
        response = requests.post(request_url, data=params, headers=headers)
        if response:
            data = response.json()
            if data['error_msg'] == 'SUCCESS':
                return data
            else:
                QMessageBox.about(self, "获取用户人脸列表结果", "获取用户人脸列表
                                  失败\n" + data['error_msg'])
```

(17) 编写函数 del_face_token()，功能是删除系统中已经存在的某个人脸信息，会同步在百度 AI 的应用程序中删除这个人脸信息。代码如下：

```
    def del_face_token(self,group,user,face_token):
        request_url = "https://aip.baidubce.com/rest/2.0/face/v3/faceset/
                       face/delete"

        params = {
            "user_id": user,
            "group_id": group,
            "face_token":face_token
        }
        access_token = self.access_token
        request_url = request_url + "?access_token=" + access_token
        headers = {'content-type': 'application/json'}
```

```
    response = requests.post(request_url, data=params, headers=headers)
    if response:
        data = response.json()
        if data['error_msg'] == 'SUCCESS':
            QMessageBox.about(self, "人脸删除结果", "人脸删除成功")
        else:
            QMessageBox.about(self, "人脸删除结果", "用户组删除失败\n"
                              + data['error_msg'])
```

(18) 编写函数 del_user()，功能是删除系统中已经存在的某个用户信息，会同步在百度AI的应用程序中删除这个用户信息。代码如下：

```
def del_user(self):
    #查询用户人脸信息(face_token)
    #获取用户组
    list = self.getlist()
    group,ret = QInputDialog.getText(self,"用户组获取","用户组信息\n"
                +str(list['result']['group_id_list']))
    group_status = 0
    if self.group_id == '':
        QMessageBox.about(self, "删除失败", "用户组不能为空\n")
        return
    for i in list['result']['group_id_list']:
        if i == group:
            group_status = 1
            break
    if group_status == 0:
        QMessageBox.about(self, "删除失败", "该用户组不存在\n")
        return
    #获取用户
    userlist = self.get_userlist(group)
    user,ret = QInputDialog.getText(self,"用户获取","用户信息\n"
               +str(userlist['result']['user_id_list']))
    user_status = 0
    if user == '':
        QMessageBox.about(self, "删除失败", "用户不能为空\n")
        return
    for i in userlist['result']['user_id_list']:
        if i == user:
            user_status = 1
            break
    if user_status == 0:
        QMessageBox.about(self, "删除失败", "该用户不存在\n")
        return
    #获取用户人脸列表
    face_list = self.user_face_list(group,user)
```

```
    for i in face_list['result']['face_list']:
        self.del_face_token(group,user,i['face_token'])
```

9.4.4 多线程操作和人脸识别

为了提高本项目的运行效率，使用多线程技术提高了人脸识别的能力。编写文件 detect.py，功能是使用多编程技术实现人脸识别，并创建 SQLite 数据库，然后将添加的用户信息和用户组信息添加到数据库中。文件 detect.py 的具体实现流程如下所示。

(1) 创建线程类 detect_thread，使用多线程技术向百度 AI 服务器传递识别信息，代码如下：

```
class detect_thread(QThread):
    transmit_data = pyqtSignal(dict)
    search_data = pyqtSignal(str)
    OK = True
    #字典用来存放签到数据
    sign_list = {}
    def __init__(self,token,group_id):
        super(detect_thread,self).__init__()#初始化操作
        self.access_token = token
        self.group_id = group_id
        self.condition = False
        self.add_status = 0
        # self.create_sqlite()

    #run 函数执行结束，代表线程结束
    def run(self):
        print("run")
        '''
        self.time = QTimer(self)
        self.time.start(500)
        self.time.timeout.connect(self.detect_face)
        '''
        while self.OK:
            if self.condition:
                self.detect_face(self.base64_image)
                self.condition = False
        print("while finish")
```

执行线程类，只会执行线程类中的函数 run()，如果需要实现新的功能，则需要重新写一个 run()函数。

(2) 编写函数 get_base64()，功能是获取传递的识别数据，将 base64 格式的图片发送给

百度 AI，以便于获取人脸信息。代码如下：

```
def get_base64(self,base64_image):
    #当窗口产生信号，调用槽函数，就把传递的数据，存放在线程的变量中
    self.base64_image = base64_image
    self.condition = True

    con = sqlite3.connect(r"stu_data.db")
    c = con.cursor()
```

(3) 编写函数 detect_face()，功能是使用多线程类向百度 AI 发送请求，并获取百度 AI 返回的人脸信息。向百度 AI 发送的是 base64 格式的图片，将人脸信息返回到文件 mywindow.py 中。代码如下：

```
def detect_face(self,base64_image):
    '''
    #对话框获取图片
    #获取一张图片(一帧画面)
    #getOpenFileName通过对话框的形式获取一个图片(.JPG)路径
    path,ret = QFileDialog.getOpenFileName(self,"open picture",".",
                "图片格式(*.jpg)")
    #把图片转换成base64编码格式
    fp = open(path,'rb')
    base64_imag = base64.b64encode(fp.read())
    print(base64_imag)
    '''
    # 摄像头获取画面
    # camera_data = self.cameravideo.read_camera()
    # 把摄像头画面转换成图片，然后设置编码base64编码格式数据
    # _, enc = cv2.imencode('.jpg', camera_data)
    # base64_image = base64.b64encode(enc.tobytes())
    # 发送请求的地址
    request_url = "https://aip.baidubce.com/rest/2.0/face/v3/detect"
    # 请求参数是一个字典，在字典中存储,百度AI要识别的图片信息，属性内容
    params = {
        "image": base64_image,  # 图片信息字符串
        "image_type": "BASE64",  # 图片信息格式
        "face_field": "gender,age,beauty,expression,face_shape,glasses,
                emotion, mask", # 请求识别人脸的属性， 各个属性在字符串中用逗号隔开
        "max_face_num": 1
    }
    # 访问令牌
    access_token = self.access_token
    # 把请求地址和访问令牌组成可用的网络请求
    request_url = request_url + "?access_token=" + access_token
```

```
    # 参数：设置请求的格式体
    headers = {'content-type': 'application/json'}
    # 发送网络 post 请求,请求百度 AI 进行人脸检测,返回检测结果
    # 发送网络请求，就会存在一定的等待时间，程序就在这里阻塞执行，所以会存在卡顿现象
    response = requests.post(request_url, data=params, headers=headers)
    if response:
        data = response.json()
        if data['error_code'] != 0:
            self.transmit_data.emit(data)
            self.search_data.emit(data['error_msg'])
            return

        if data['result']['face_num'] > 0:
            #data 是请求数据的结果，需要进行解析，单独拿出所需的数据内容，分开
            self.transmit_data.emit(dict(data))
            self.face_search(self.group_id)
```

(4) 编写函数 face_search()，功能是使用百度 AI 实现人脸识别功能。代码如下：

```
def face_search(self,group_id):
    request_url = "https://aip.baidubce.com/rest/2.0/face/v3/search"
    params = {
        "image": self.base64_image,
        "image_type": "BASE64",
        "group_id_list": group_id,
    }
    access_token = self.access_token
    request_url = request_url + "?access_token=" + access_token
    headers = {'content-type': 'application/json'}
    response = requests.post(request_url, data=params, headers=headers)
    if response:
        data = response.json()
        if data['error_code'] == 0:
            if data['result']['user_list'][0]['score'] > 90:
                #存储要保存的签到数据，方便显示
                del(data['result']['user_list'][0]['score'])
                datetime = QDateTime.currentDateTime()
                datetime = datetime.toString()
                data['result']['user_list'][0]['datetime'] = datetime
                key = data['result']['user_list'][0]['group_id']
                      +data['result']['user_list'][0]['user_id']
                if key not in self.sign_list.keys():
                    self.sign_list[key] = data['result']['user_list'][0]
                self.search_data.emit("学生签到成功\n 学生信息："
                                +data['result']['user_list'][0]['user_info'])
                stu_data = data['result']['user_list'][0]['user_info']
                info = stu_data.split('\n')
```

```
                _, info_name = info[0].split(': ')
                _, info_class = info[1].split(': ')
                id = data['result']['user_list'][0]['user_id']
                # self.add_sqlite(id, info_name, info_class, datetime)
                self.search_sqlite(id)
                search_id=0
                for i in self.values:
                    search_id = i[0]
                if search_id == id:
                    self.update_sqlite(id,info_name,info_class,datetime)
                else:
                    self.add_sqlite(id,info_name,info_class,datetime)
            else:
                self.search_data.emit("学生签到失败，找不到对应学生")
```

(5) 编写函数 create_sqlite(self)创建数据库 stu_data.db 和表 student，代码如下：

```
def create_sqlite(self):
    con = sqlite3.connect(r"stu_data.db")
    c = con.cursor()
    c.execute("create table student(id primary key ,name ,stu_class,datetime)")
    print("创建成功")
```

(6) 编写函数 add_sqlite()，功能是将学生信息添加到数据库中，代码如下：

```
def add_sqlite(self,id,name,stu_class,datetime):
    con = sqlite3.connect(r"stu_data.db")
    c = con.cursor()
    value = (id,name,stu_class,datetime)
    sql = "insert into student(id,name,stu_class,datetime) values(?,?,?,?)"
    c.execute(sql,value)
    print("添加成功")
    # 提交
    con.commit()
```

(7) 编写函数 update_sqlite()，功能是更新数据库中某个学生的信息，代码如下：

```
def update_sqlite(self,id,name,stu_class,datetime):
    con = sqlite3.connect(r"stu_data.db")
    c = con.cursor()
    # value = (name,stu_class,datetime,id)
    sql = "update student set name=?,stu_class=?,datetime=? where id =?"
    c.execute(sql,(name,stu_class,datetime,id))
    con.commit()
    print("更新成功")
```

(8) 编写函数 search_sqlite(self,id)，功能是查询数据库中某个学生的信息，代码如下：

```
    def search_sqlite(self,id):
        con = sqlite3.connect(r"stu_data.db")
        c = con.cursor()
        sql = "select * from student where id=?"
        self.values = c.execute(sql,(id,))
```

9.4.5 导出打卡签到信息

某个学生成功打卡签到后，可以导出本次签到信息，将信息保存到本地文件中。

(1) 首先使用 Qt Designer 设计 UI 界面文件 sign_indata.ui，效果如图 9-6 所示。

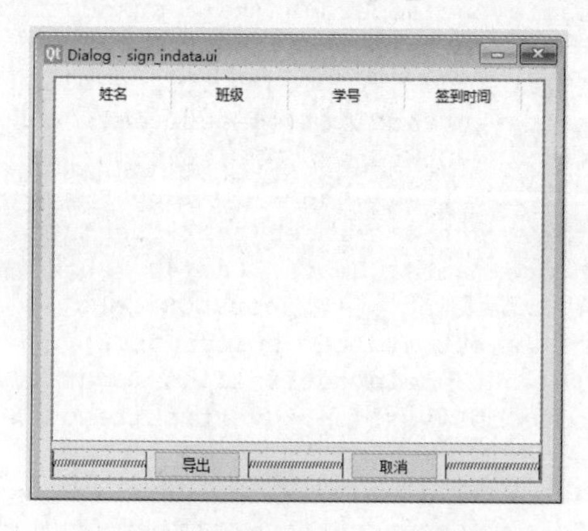

图 9-6 UI 界面文件 sign_indata.ui

(2) 将 Qt Designer 文件 sign_indata.ui 转换为 Python 文件 sign_indata.py，代码如下：

```
class Ui_Dialog(object):
    def setupUi(self, Dialog):
        Dialog.setObjectName("Dialog")
        Dialog.resize(440, 353)
        self.gridLayout = QtWidgets.QGridLayout(Dialog)
        self.gridLayout.setObjectName("gridLayout")
        self.verticalLayout = QtWidgets.QVBoxLayout()
        self.verticalLayout.setObjectName("verticalLayout")
        self.tableWidget = QtWidgets.QTableWidget(Dialog)
        self.tableWidget.setMinimumSize(QtCore.QSize(0, 0))
        self.tableWidget.setObjectName("tableWidget")
        self.tableWidget.setColumnCount(4)
        self.tableWidget.setRowCount(0)
        item = QtWidgets.QTableWidgetItem()
        self.tableWidget.setHorizontalHeaderItem(0, item)
        item = QtWidgets.QTableWidgetItem()
        self.tableWidget.setHorizontalHeaderItem(1, item)
```

```
        item = QtWidgets.QTableWidgetItem()
        self.tableWidget.setHorizontalHeaderItem(2, item)
        item = QtWidgets.QTableWidgetItem()
        self.tableWidget.setHorizontalHeaderItem(3, item)
        self.verticalLayout.addWidget(self.tableWidget)
        self.horizontalLayout = QtWidgets.QHBoxLayout()
        self.horizontalLayout.setObjectName("horizontalLayout")
        spacerItem = QtWidgets.QSpacerItem(40, 20, QtWidgets.QSizePolicy.
                    Expanding, QtWidgets.QSizePolicy.Minimum)
        self.horizontalLayout.addItem(spacerItem)
        self.pushButton = QtWidgets.QPushButton(Dialog)
        self.pushButton.setObjectName("pushButton")
        self.horizontalLayout.addWidget(self.pushButton)
        spacerItem1 = QtWidgets.QSpacerItem(40, 20, QtWidgets.QSizePolicy.
                    Expanding, QtWidgets.QSizePolicy.Minimum)
        self.horizontalLayout.addItem(spacerItem1)
        self.pushButton_2 = QtWidgets.QPushButton(Dialog)
        self.pushButton_2.setObjectName("pushButton_2")
        self.horizontalLayout.addWidget(self.pushButton_2)
        spacerItem2 = QtWidgets.QSpacerItem(40, 20, QtWidgets.QSizePolicy.
                    Expanding, QtWidgets.QSizePolicy.Minimum)
        self.horizontalLayout.addItem(spacerItem2)
        self.verticalLayout.addLayout(self.horizontalLayout)
        self.gridLayout.addLayout(self.verticalLayout, 0, 0, 1, 1)

        self.retranslateUi(Dialog)
        QtCore.QMetaObject.connectSlotsByName(Dialog)

    def retranslateUi(self, Dialog):
        _translate = QtCore.QCoreApplication.translate
        Dialog.setWindowTitle(_translate("Dialog", "导出"))
        item = self.tableWidget.horizontalHeaderItem(0)
        item.setText(_translate("Dialog", "姓名"))
        item = self.tableWidget.horizontalHeaderItem(1)
        item.setText(_translate("Dialog", "班级"))
        item = self.tableWidget.horizontalHeaderItem(2)
        item.setText(_translate("Dialog", "学号"))
        item = self.tableWidget.horizontalHeaderItem(3)
        item.setText(_translate("Dialog", "签到时间"))
        self.pushButton.setText(_translate("Dialog", "导出"))
        self.pushButton_2.setText(_translate("Dialog", "取消"))
```

(3) 编写文件sign_indata.py，功能是导出某次打卡签到的信息，代码如下：

```
class sign_data(Ui_Dialog,QDialog):
    def __init__(self, signdata,parent=None):
        super(sign_data, self).__init__(parent)
```

```
        self.setupUi(self)#创建界面内容
        #设置窗口内容不能被修改
        self.signdata = signdata
        self.tableWidget.setEditTriggers(QAbstractItemView.NoEditTriggers)
        for i in signdata.values():
            info = i['user_info'].split('\n')
            rowcount = self.tableWidget.rowCount()
            self.tableWidget.insertRow(rowcount)
            info_name = info[0].split(': ')
            self.tableWidget.setItem(rowcount, 0, QTableWidgetItem(info_name[1]))
            info_class = info[1].split(': ')
            self.tableWidget.setItem(rowcount, 1, QTableWidgetItem(info_class[1]))
            self.tableWidget.setItem(rowcount, 2, QTableWidgetItem(i['user_id']))
            self.tableWidget.setItem(rowcount, 3, QTableWidgetItem(i['datetime']))
        #导出按钮
        self.pushButton.clicked.connect(self.save_data)
        #取消按钮
        self.pushButton_2.clicked.connect(self.close_window)
    def close_window(self):
        self.reject()

    def save_data(self):
        #打开对话框，获取要导出的数据文件名
        filename,ret = QFileDialog.getSaveFileName(self,"导出数据",".","TXT(*.txt)")
        f = open(filename,"w")
        for i in self.signdata.values():
            info = i['user_info'].split('\n')
            _,info_name = info[0].split(': ')
            _,info_class = info[1].split(': ')
            f.write(str(info_name+" "+info_class+" "+i['user_id']+" "
                    +i['datetime'] ))
        f.close()
        self.accept()
```

9.5 调试运行

运行文件 main.py，系统主界面的执行效果如图 9-7 所示。

选择菜单栏中“用户组”|“添加用户组”命令，首先弹出“添加用户组”对话框，输入合法的用户组名字，然后单击 OK 按钮，会向系统数据库中添加这个用户组的信息，并且在百度 AI 的应用中也会同步创建这个用户组，如图 9-8 所示。

扫码观看本节视频讲解

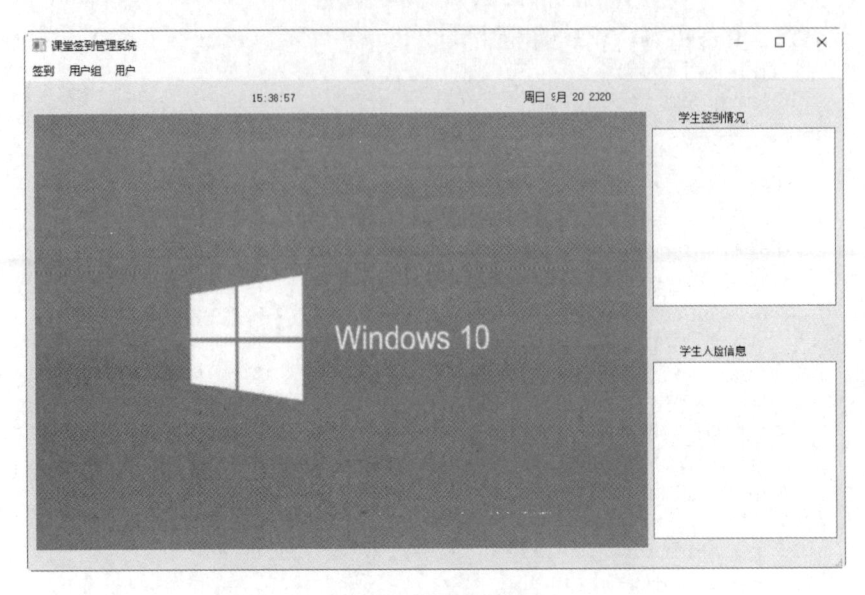

图 9-7　系统主界面

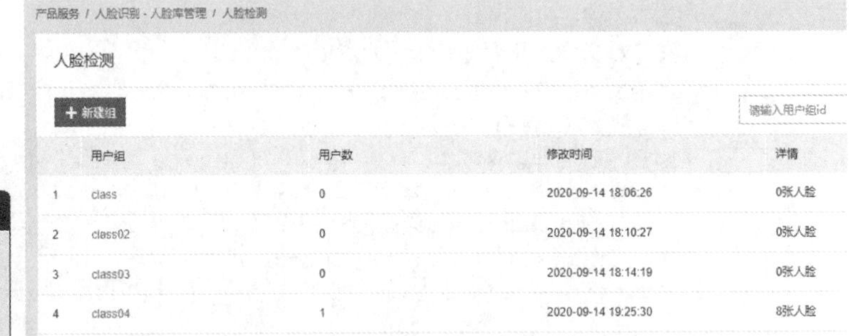

“添加用户组”对话框　　　　同步在百度 AI 应用中创建这个用户组

图 9-8　新建用户组操作

选择菜单栏中“用户”|“添加用户”命令，弹出“添加用户”对话框，首先选择这个用户所属的用户组，然后输入这个用户的姓名、学号和班级信息，最后单击“确定”按钮，向系统数据库中添加这个用户的信息，并且在百度 AI 的应用中同步添加这个用户的信息，如图 9-9 所示。

选择菜单栏中“签到”|“启动签到”命令，弹出“请输入所在用户组”对话框，输入用户组并单击 OK 按钮后会打开摄像头实现人脸识别功能，如果识别成功则自动进行打卡签到，如图 9-10 所示。

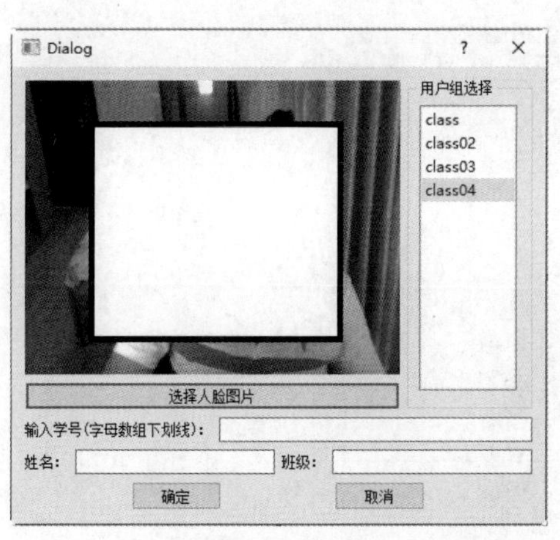

“添加用户”对话框

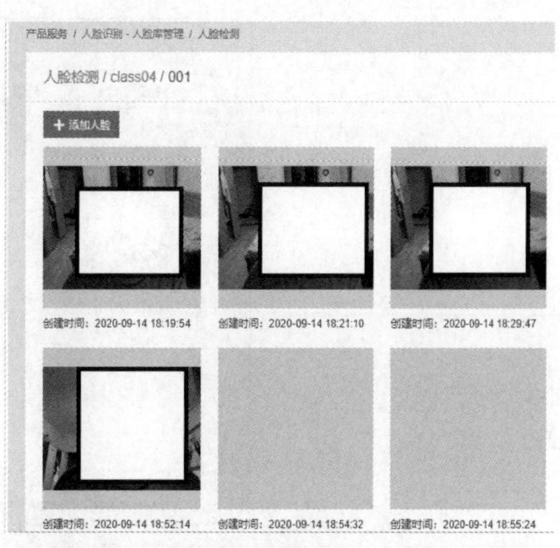

同步在百度 AI 应用中添加这个用户的信息

图 9-9 添加用户操作

请输入所在用户组

用户组信息

['class', 'class02', 'class03', 'class04']

OK Cancel

输入所在用户组

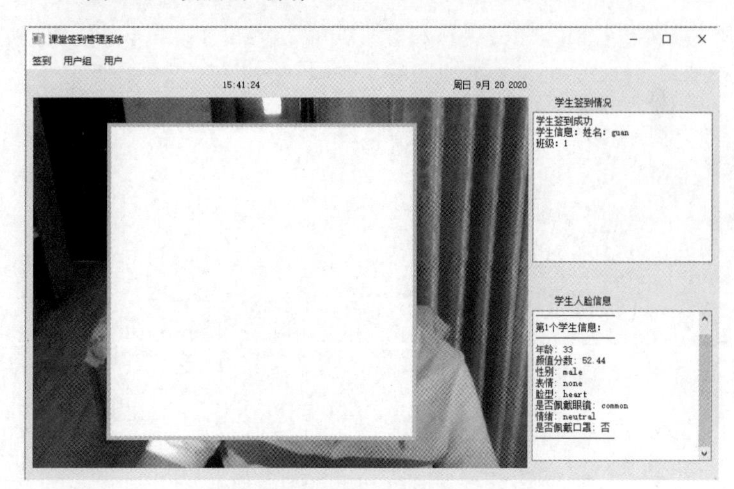

自动打卡签到

图 9-10 “启动签到”命令

选择菜单栏中“签到”|“关闭签到”命令，在弹出的对话框中单击“导出”按钮，会将本次打卡签到信息导出到文本文件中，如图 9-11 所示。

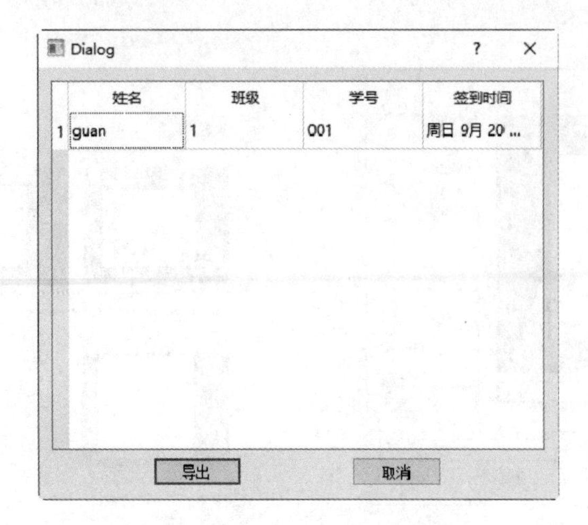

图 9-11　导出本次打卡签到信息

第10章 基于深度学习的AI人脸识别系统(Flask+OpenCV-Python+Keras+Sklearn实现)

近年来，随着人工智能技术的飞速发展，机器学习和深度学习技术已经摆在了人们的面前，一时间成为程序员们的学习热点。在本章的内容中，将详细介绍使用深度学习技术开发一个人脸识别系统的知识，详细讲解了使用OpenCV-Python+ Keras+Sklearn实现一个大型人工智能项目的过程。

10.1 系统需求分析

本章将详细讲解使用人工智能技术实现一个大型人脸识别项目的过程。在本节将首先讲解本项目的需求分析，为步入后面知识的学习打下基础。

扫码观看本节视频讲解

10.1.1 系统功能分析

本项目是一个人工智能版的人脸识别系统，使用深度学习技术实现。本项目的具体功能模块如下所示。

(1) 采集样本照片。

调用本地电脑摄像头采集照片作为样本，设置快捷键进行采集和取样。一次性可以采集无数个照片，采集的样本照片越多，后面的人脸识别成功率越高。

(2) 图片处理。

处理采集到的原始样本照片，将采集到的原始图像转换为标准数据文件，这样便于被后面的深度学习模块所用。

(3) 深度学习。

使用处理后的图片创建深度学习模型，实现学习训练，将训练结果保存为“.h5”文件。

(4) 人脸识别。

根据训练所得的模型实现人脸识别功能，既可以识别摄像头中的图片，也可以识别 Flask Web 中的上传照片。

10.1.2 实现流程分析

实现本项目的具体流程如图 10-1 所示。

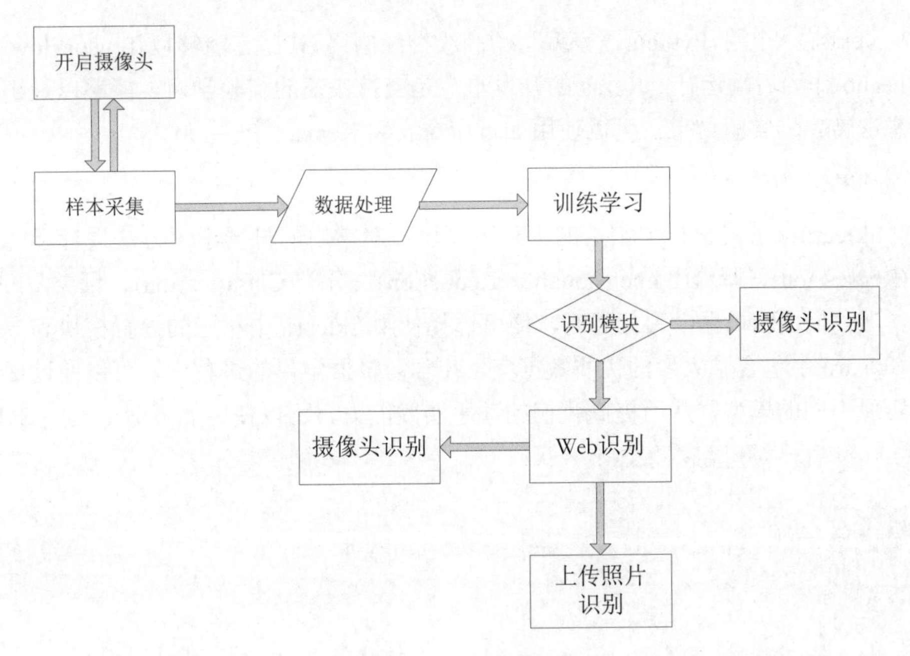

图 10-1 实现流程

10.1.3 技术分析

本人脸识别系统是一个综合性的项目，主要用到了如下所示的框架。

(1) Flask：著名的 Python Web 开发框架。

(2) OpenCV-Python：著名的图像处理框架 OpenCV 的 Python 接口。OpenCV 是一个基于 BSD 许可(开源)发行的跨平台计算机视觉库，可以运行在 Linux、Windows、Android 和 Mac OS 操作系统上。它轻量级而且高效——由一系列 C 函数和少量 C++类构成，同时提供了 Python、Ruby、MATLAB 等语言的接口，实现了图像处理和计算机视觉方面的很多通用算法。OpenCV 用 C++语言编写，它的主要接口也是 C++语言，但是依然保留了大量的 C 语言接口。

可以使用如下命令安装 OpenCV-Python：

```
pip install opencv-python
```

在安装 OpenCV-Python 时需要安装对应的依赖库，例如常用的 NumPy 等。如果安装 OpenCV-Python 失败，可以下载对应的“.whl”文件，然后通过如下命令进行安装：

```
pip install “.whl”文件
```

(3) Keras：一个用 Python 语言编写的高级神经网络 API，它能够以 TensorFlow、CNTK 或者 Theano 作为后端运行。Keras 的开发重点是支持快速的实验学习，能够以最小的时延把你的想法转换为实验结果。可以使用如下命令安装 Keras：

```
pip install keras
```

(4) Sklearn：机器学习中常用的第三方模块，对常用的机器学习方法进行了封装，包括回归(Regression)、降维(Dimensionality Reduction)、分类(Classfication)、聚类(Clustering)等方法。当我们面临机器学习问题时，便可选择使用 Sklearn 中相应的内置模块和方法来实现。在 Sklearn 中包含了大量的优质数据集，在学习机器学习的过程中，可以通过使用这些数据集实现不同的模型，从而提高我们的动手实践能力。可以使用如下命令安装 Sklearn：

```
pip install sklearn
```

注意： 在安装 Sklearn 之前，需要先安装 NumPy 和 Scipy。

10.2 照片样本采集

在进行人脸识别之前，需要先采集一个照片作为样本。在本节的内容中，将详细讲解使用摄像头采集样本照片的过程。

扫码观看本节视频讲解

编写文件 getCameraPics.py，基于摄像头采集视频流中的数据，将截取的人脸图片作为样本照片并存储起来。文件 getCameraPics.py 的具体实现代码如下所示：

```
import os
import cv2

#python2 运行时加上
# reload(sys)
# sys.setdefaultencoding('utf-8')

def cameraAutoForPictures(saveDir='data/'):
    '''
    调用电脑摄像头来自动获取图片
    '''
    if not os.path.exists(saveDir):
        os.makedirs(saveDir)
```

```
    count=1
    cap=cv2.VideoCapture(0)
    width,height,w=640,480,360
    cap.set(cv2.CAP_PROP_FRAME_WIDTH,width)
    cap.set(cv2.CAP_PROP_FRAME_HEIGHT,height)
    crop_w_start=(width-w)//2
    crop_h_start=(height-w)//2
    print('width: ',width)
    print('height: ',height)
    while True:
        ret,frame=cap.read()
        frame=frame[crop_h_start:crop_h_start+w,crop_w_start:crop_w_start+w]
        frame=cv2.flip(frame,1,dst=None)
        cv2.imshow("capture", frame)
        action=cv2.waitKey(1) & 0xFF
        if action==ord('c'):
            saveDir=input(u"请输入新的存储目录：")
            if not os.path.exists(saveDir):
                os.makedirs(saveDir)
        elif action==ord('p'):
            cv2.imwrite("%s/%d.jpg" % (saveDir,count),cv2.resize(frame, (224, 224),
                        interpolation=cv2.INTER_AREA))
            print(u"%s: %d 张图片" % (saveDir,count))
            count+=1
        if action==ord('q'):
            break
    cap.release()
    cv2.destroyAllWindows()


if __name__=='__main__':
    cameraAutoForPictures(saveDir='data/guanxijing/')
```

通过上述代码，启动摄像头后需要借助键盘输入操作来完成图片的获取工作，其中键盘按键 c(change)表示设置一个存储样本照片的目录，按键 p(photo)表示执行截图操作，按键 q(quit)表示退出拍摄。运行代码后会打开本地电脑中的摄像头，如图 10-2 所示。按键盘中的 p 键，会截取摄像头中的屏幕照片，并将照片保存起来，上述代码设置的保存目录是 guanxijing。

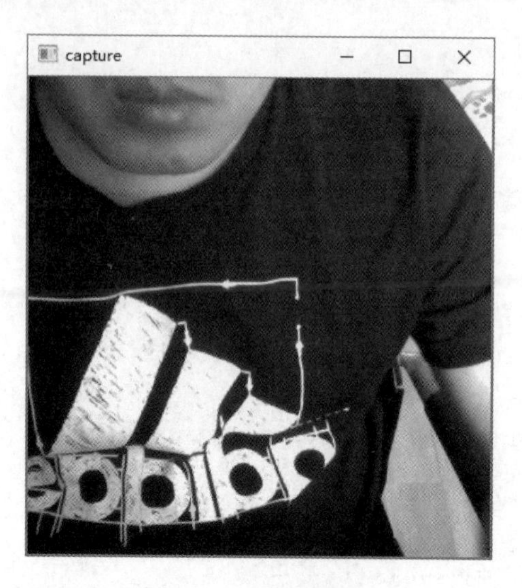

图 10-2　开启摄像头

10.3　深度学习和训练

在尽可能多地采集样本照片后，需要将采集到的数据进行分析处理，然后使用人工智能技术实现深度学习训练，再将训练结果保存为数据模型文件，根据数据模型文件可以实现人脸识别功能。

扫码观看本节视频讲解

10.3.1　原始图像预处理

编写文件 dataHelper.py，实现原始图像数据的预处理工作，将原始图像转换为标准数据文件。文件 dataHelper.py 的具体实现代码如下所示：

```
import os
import cv2
import time

def readAllImg(path,*suffix):
    '''
    基于后缀读取文件
    '''
    try:
        s=os.listdir(path)
        resultArray = []
```

```
        fileName = os.path.basename(path)
        resultArray.append(fileName)
        for i in s:
            if endwith(i, suffix):
                document = os.path.join(path, i)
                img = cv2.imread(document)
                resultArray.append(img)
    except IOError:
        print("Error")

    else:
        print("读取成功")
        return resultArray


def endwith(s,*endstring):
    '''
    对字符串的后缀进行匹配
    '''
    resultArray = map(s.endswith,endstring)
    if True in resultArray:
        return True
    else:
        return False

def readPicSaveFace(sourcePath,objectPath,*suffix):
    '''
    图片标准化与存储
    '''
    if not os.path.exists(objectPath):
        os.makedirs(objectPath)
    try:
        resultArray=readAllImg(sourcePath,*suffix)
        count=1
        face_cascade=cv2.CascadeClassifier('config/haarcascade_
                    frontalface_alt.xml')
        for i in resultArray:
            if type(i)!=str:
              gray=cv2.cvtColor(i, cv2.COLOR_BGR2GRAY)
              faces=face_cascade.detectMultiScale(gray, 1.3, 5)
              for (x,y,w,h) in faces:
                listStr=[str(int(time.time())),str(count)]
                fileName=''.join(listStr)
                f=cv2.resize(gray[y:(y+h),x:(x+w)],(200, 200))
                cv2.imwrite(objectPath+os.sep+'%s.jpg' % fileName, f)
                count+=1
    except Exception as e:
```

```
        print("Exception: ",e)
    else:
        print('Read  '+str(count-1)+' Faces to Destination '+objectPath)

if __name__ == '__main__':
    print('dataProcessing!!!')
    readPicSaveFace('data/guanxijing/','dataset/guanxijing/','.jpg',
                    '.JPG','png','PNG','tiff')

readPicSaveFace('data/KA/','dataset/KA/','.jpg','.JPG','png','PNG','tiff')
```

如果需要处理多个人的样本照片，需要在__main__后面添加多个对应的处理目录。运行上述文件后，会在 dataset 目录下得到处理后的照片。

10.3.2 构建人脸识别模块

编写文件 faceRegnigtionModel.py，功能是通过深度学习和训练构建人脸识别模块，将训练后得到的模型保存到本地，默认保存为 face.h5。文件 faceRegnigtionModel.py 的具体实现流程如下所示。

(1) 引入深度学习和机器学习框架，对应的实现代码如下所示：

```
import os
import cv2
import random
import numpy as np
from keras.utils import np_utils
from keras.models import Sequential,load_model
from sklearn.model_selection import train_test_split
from keras.layers import
Dense,Activation,Convolution2D,MaxPooling2D,Flatten,Dropout
```

(2) 编写类 DataSet，功能是保存和读取格式化后的训练数据。对应的实现代码如下所示：

```
class DataSet(object):
    def __init__(self,path):
        '''
        初始化
        '''
        self.num_classes=None
        self.X_train=None
        self.X_test=None
        self.Y_train=None
        self.Y_test=None
```

```
    self.img_size=128
    self.extract_data(path)
```

(3) 编写函数 extract_data()抽取数据，使用机器学习 Sklearn 中的函数 train_test_split()将原始数据集按照一定比例划分训练集和测试集对模型进行训练。通过函数 reshape()将图片转换成指定的尺寸和灰度，通过函数 astype()将图片转换为 float32 数据类型。对应的实现代码如下所示：

```
def extract_data(self,path):
    imgs,labels,counter=read_file(path)
    X_train,X_test,y_train,y_test=train_test_split(imgs,labels,
                    test_size=0.2,random_state=random.randint(0, 100))
    X_train=X_train.reshape(X_train.shape[0], 1, self.img_size,
                            self.img_size)/255.0
    X_test=X_test.reshape(X_test.shape[0], 1, self.img_size,
                          self.img_size)/255.0
    X_train=X_train.astype('float32')
    X_test=X_test.astype('float32')
    Y_train=np_utils.to_categorical(y_train, num_classes=counter)
    Y_test=np_utils.to_categorical(y_test, num_classes=counter)
    self.X_train=X_train
    self.X_test=X_test
    self.Y_train=Y_train
    self.Y_test=Y_test
    self.num_classes=counter
```

函数 train_test_split()的原型如下：

```
train_test_split(train_data,train_target,test_size,random_state)
```

函数 train_test_split()各个参数的具体说明如下所示。

- train_data：表示被划分的样本特征集。
- train_target：表示划分的样本的标签(索引值)。
- test_size：表示将样本按比例划分，返回的第一个参数值为 train_data*test_size。
- random_state：表示随机种子。当为整数的时候，不管循环多少次 X_train 与第一次一样的。其值不能是小数，当 random_state 的值发生改变的时候，其返回值也会发生改变。

(4) 编写函数 check(self)实现数据校验，打印输出图片的基本信息。对应的实现代码如下所示：

```
def check(self):
    '''
    校验
```

```
        '''
        print('num of dim:', self.X_test.ndim)
        print('shape:', self.X_test.shape)
        print('size:', self.X_test.size)
        print('num of dim:', self.X_train.ndim)
        print('shape:', self.X_train.shape)
        print('size:', self.X_train.size)
```

(5) 编写函数 endwith()，功能是对字符串的后续和标签进行匹配。对应的实现代码如下所示：

```
def endwith(s,*endstring):
   resultArray = map(s.endswith,endstring)
   if True in resultArray:
      return True
   else:
      return False
```

(6) 编写函数 read_file(path)读取指定路径的图片信息，对应的实现代码如下所示：

```
def read_file(path):
   img_list=[]
   label_list=[]
   dir_counter=0
   IMG_SIZE=128
   for child_dir in os.listdir(path):
      child_path=os.path.join(path, child_dir)
      for dir_image in os.listdir(child_path):
         if endwith(dir_image,'jpg'):
            img=cv2.imread(os.path.join(child_path, dir_image))
            resized_img=cv2.resize(img, (IMG_SIZE, IMG_SIZE))
            recolored_img=cv2.cvtColor(resized_img,cv2.COLOR_BGR2GRAY)
            img_list.append(recolored_img)
            label_list.append(dir_counter)
      dir_counter+=1
   img_list=np.array(img_list)
   return img_list,label_list,dir_counter
```

(7) 编写函数 read_name_list(path)读取训练数据集，对应的实现代码如下所示：

```
def read_name_list(path):
   name_list=[]
   for child_dir in os.listdir(path):
      name_list.append(child_dir)
   return name_list
```

(8) 编写类 Model，创建一个基于 CNN 的人脸识别模型，开始构建数据模型并进行训练。对应的实现代码如下所示：

```
class Model(object):
    '''
    人脸识别模型
    '''
    FILE_PATH="face.h5"
    IMAGE_SIZE=128


    def __init__(self):
        self.model=None


    def read_trainData(self,dataset):
        self.dataset=dataset


    def build_model(self):
        self.model = Sequential()
        self.model.add(
            Convolution2D(
                filters=32,
                kernel_size=(5, 5),
                padding='same',
                dim_ordering='th',
                input_shape=self.dataset.X_train.shape[1:]
            )
        )
        self.model.add(Activation('relu'))
        self.model.add(
            MaxPooling2D(
                pool_size=(2, 2),
                strides=(2, 2),
                padding='same'
            )
        )
        self.model.add(Convolution2D(filters=64, kernel_size=(5,5), padding='same'))
        self.model.add(Activation('relu'))
        self.model.add(MaxPooling2D(pool_size=(2,2), strides=(2,2), padding='same'))
        self.model.add(Flatten())
        self.model.add(Dense(1024))
        self.model.add(Activation('relu'))
        self.model.add(Dense(self.dataset.num_classes))
        self.model.add(Activation('softmax'))
        self.model.summary()


    def train_model(self):
```

```
        self.model.compile(
            optimizer='adam',
            loss='categorical_crossentropy',
            metrics=['accuracy'])
        self.model.fit(self.dataset.X_train,self.dataset.Y_train,
                       epochs=10,batch_size=10)


    def evaluate_model(self):
        print('\nTesting---------------')
        loss, accuracy = self.model.evaluate(self.dataset.X_test, self.dataset.Y_test)
        print('test loss;', loss)
        print('test accuracy:', accuracy)


    def save(self, file_path=FILE_PATH):
        print('Model Saved Finished!!!')
        self.model.save(file_path)


    def load(self, file_path=FILE_PATH):
        print('Model Loaded Successful!!!')
        self.model = load_model(file_path)


    def predict(self,img):
        img=img.reshape((1, 1, self.IMAGE_SIZE, self.IMAGE_SIZE))
        img=img.astype('float32')
        img=img/255.0
        result=self.model.predict_proba(img)
        max_index=np.argmax(result)
        return max_index,result[0][max_index]
```

(9) 调用上面的函数打印输出模型训练和评估结果，对应的实现代码如下所示：

```
if __name__ == '__main__':
    dataset=DataSet('dataset/')
    model=Model()
    model.read_trainData(dataset)
    model.build_model()
    model.train_model()
    model.evaluate_model()
    model.save()
```

10.4 人脸识别

在使用人工智能技术实现深度学习训练后，可以生成一个数据模型文件，通过调用这个数据模型文件可以实现人脸识别功能。例如在下面的实例文件 cameraDemo.py 中，通过 OpenCV-Python 直接调用摄像头实现实时人脸识别功能。

扫码观看本节视频讲解

```
import os
import cv2
from faceRegnigtionModel import Model

threshold=0.7  #如果模型认为概率高于 70%则显示为模型中已有的人物

def read_name_list(path):
    '''
    读取训练数据集
    '''
    name_list=[]
    for child_dir in os.listdir(path):
        name_list.append(child_dir)
    return name_list

class Camera_reader(object):
    def __init__(self):
        self.model=Model()
        self.model.load()
        self.img_size=128

    def build_camera(self):
        '''
        调用摄像头来实时人脸识别
        '''
        face_cascade = cv2.CascadeClassifier('config/haarcascade_
                        frontalface_alt.xml')
        name_list=read_name_list('dataset/')
        cameraCapture=cv2.VideoCapture(0)
        success, frame=cameraCapture.read()
        while success and cv2.waitKey(1)==-1:
            success,frame=cameraCapture.read()
            gray=cv2.cvtColor(frame, cv2.COLOR_BGR2GRAY)
            faces=face_cascade.detectMultiScale(gray, 1.3, 5)
            for (x,y,w,h) in faces:
                ROI=gray[x:x+w,y:y+h]
                ROI=cv2.resize(ROI, (self.img_size, self.img_size),
                                interpolation=cv2.INTER_LINEAR)
```

```
                label,prob=self.model.predict(ROI)
                if prob>threshold:
                    show_name=name_list[label]
                else:
                    show_name="Stranger"
                cv2.putText(frame, show_name, (x,y-20),
                        cv2.FONT_HERSHEY_SIMPLEX,1,255,2)
                frame=cv2.rectangle(frame,(x,y), (x+w,y+h),(255,0,0),2)
            cv2.imshow("Camera", frame)
        else:
            cameraCapture.release()
            cv2.destroyAllWindows()

if __name__ == '__main__':
    camera=Camera_reader()
    camera.build_camera()
```

执行代码后会开启摄像头并识别摄像头中的人物，如图 10-3 所示。

图 10-3　执行效果

10.5　Flask Web 人脸识别接口

我们可以将前面实现的数据模型和人脸识别功能迁移到 Web 项目中。在本节的内容中，将详细讲解在 Flask Web 中实现人脸识别功能的过程。

扫码观看本节视频讲解

10.5.1　导入库文件

编写文件 main.py，实现 Flask 项目的主程序功能。首先导入需要的人脸识别库和 Flask

库，具体实现代码如下所示：

```
from flask_uploads import UploadSet, IMAGES, configure_uploads
from flask import redirect, url_for, render_template
import os
import cv2
import time
import numpy as np
from flask import Flask
from flask import request
from faceRegnigtionModel import Model
from cameraDemo import Camera_reader
```

10.5.2 识别上传照片

在文件 main.py 中设置 Flask 项目的名字，设置上传文件的保存目录。通过链接/upload 显示上传表单页面，通过链接/photo/<name>显示上传的照片，并在页面中调用函数 detectOnePicture(path)显示识别结果。具体实现代码如下所示：

```
app = Flask(__name__)
app.config['UPLOADED_PHOTO_DEST'] =
os.path.dirname(os.path.abspath(__file__))
app.config['UPLOADED_PHOTO_ALLOW'] = IMAGES
def dest(name):
    return '{}/{}'.format(app.config.UPLOADED_PHOTO_DEST, name)
photos = UploadSet('PHOTO')


configure_uploads(app, photos)
@app.route('/upload', methods=['POST', 'GET'])
def upload():
    if request.method == 'POST' and 'photo' in request.files:
        filename = photos.save(request.files['photo'])
        return redirect(url_for('show', name=filename))
    return render_template('upload.html')

@app.route('/photo/<name>')
def show(name):
    if name is None:
        print('出错了！')
    url = photos.url(name)

def detectOnePicture(path):
    '''
    单图识别
    '''
```

```
    model=Model()
    model.load()
    img=cv2.imread(path)
    img=cv2.resize(img,(128,128))
    img=cv2.cvtColor(img, cv2.COLOR_BGR2GRAY)
    picType,prob=model.predict(img)
    if picType!=-1:
        name_list=read_name_list('dataset/')
        print(name_list[picType],prob)
        res=u"识别为： "+name_list[picType]+u"的概率为： "+str(prob)
    else:
        res=u"抱歉，未识别出该人！请尝试增加数据量来训练模型！"
    return res

    if request.method=="GET":
        picture=name
    start_time=time.time()
    res=detectOnePicture(picture)
    end_time=time.time()
    execute_time=str(round(end_time-start_time,2))
    tsg=u' 总耗时为： %s 秒' % execute_time
    return render_template('show.html', url=url,
name=name,xinxi=res,shijian=tsg)

def endwith(s,*endstring):
    '''
    对字符串的后缀进行匹配
    '''
    resultArray=map(s.endswith,endstring)
    if True in resultArray:
        return True
    else:
        return False


def read_file(path):
    '''
    图片读取
    '''
    img_list=[]
    label_list=[]
    dir_counter=0
    IMG_SIZE=128
    for child_dir in os.listdir(path):
        child_path=os.path.join(path, child_dir)
        for dir_image in os.listdir(child_path):
            if endwith(dir_image,'jpg'):
                img=cv2.imread(os.path.join(child_path, dir_image))
                resized_img=cv2.resize(img, (IMG_SIZE, IMG_SIZE))
```

```
            recolored_img=cv2.cvtColor(resized_img,cv2.COLOR_BGR2GRAY)
            img_list.append(recolored_img)
            label_list.append(dir_counter)
        dir_counter+=1
    img_list=np.array(img_list)
    return img_list,label_list,dir_counter


def read_name_list(path):
    '''
    读取训练数据集
    '''
    name_list=[]
    for child_dir in os.listdir(path):
        name_list.append(child_dir)
    return name_list


def detectOnePicture(path):
    '''
    单图识别
    '''
    model=Model()
    model.load()
    img=cv2.imread(path)
    img=cv2.resize(img,(128,128))
    img=cv2.cvtColor(img, cv2.COLOR_BGR2GRAY)
    picType,prob=model.predict(img)
    if picType!=-1:
        name_list=read_name_list('dataset/')
        print(name_list[picType],prob)
        res=u"识别为: "+name_list[picType]+u"的概率为: "+str(prob)
    else:
        res=u"抱歉，未识别出该人！请尝试增加数据量来训练模型！"
    return res
```

10.5.3 在线识别

设置 Web 首页，显示一个“打开摄像头识别”链接，单击链接后调用摄像头实现在线识别功能。具体实现代码如下所示：

```
@app.route("/")
def init():
      return render_template("index.html",title = 'Home')

@app.route("/she/")
def she():
```

```
        camera = Camera_reader()
        camera.build_camera()
        return render_template("index.html", title='Home')

if __name__ == "__main__":
    print('faceRegnitionDemo')
    app.run(debug=True)
```

执行代码后将在主页显示“打开摄像头识别”链接，单击链接后会实现在线人脸识别功能，如图 10-4 所示。

127.0.0.1:5000

打开摄像头识别

系统主页

单击链接后启动摄像头

图 10-4　在线人脸识别

输入“http://127.0.0.1:5000/upload”后显示图片上传页面，上传照片并单击“提交”按钮后会显示上传照片和识别结果，如图 10-5 所示。

127.0.0.1:5000/upload

选择文件 未选择任何文件 提交

上传图片表单页面

显示识别结果

图 10-5　识别上传照片

第11章

AI考勤管理系统(face-recognition+Matplotlib+Django+Scikit-Learn+dlib实现)

现在已经进入了一个AI人工智能飞速发展的时代，在商业办公领域，考勤打卡应用已经实现了无纸化处理。在本章的内容中，将详细介绍使用Scikit-Learn技术开发一个AI考勤打卡系统的过程，详细讲解使用face-recognition+Matplotlib+Django+Scikit-Learn+dlib实现一个大型人工智能项目的方法。

11.1 背景介绍

随着企业人事管理的日趋复杂和企业人员的增多，企业的考勤管理变得越来越复杂，有一个比较完善的考勤管理系统变得非常重要。这时使用计算机管理方式代替以前手工处理的工作，应用计算机技术和通信技术建立一个高效率的、无差错的考勤管理系统，能够有效地帮助企业实现“公正考勤，高效薪资”，使企业的管理水平登上一个新的台阶。企业职工考勤管理系统，可用于各部门的职工考勤管理、查询、更新与维护，使用方便，易用性强，图形界面清晰明了，解决目前员工出勤管理问题，实现员工出勤信息和缺勤信息对企业领导透明，使管理人员及时把握员工的情况，及时与员工沟通，提高生产效率。

扫码观看本节视频讲解

11.2 系统需求分析

需求分析是介于系统分析和软件设计阶段之间的桥梁，好的需求分析是项目成功的基石。一方面，需求分析以系统规格说明和项目规划作为分析活动的基本出发点，并从软件角度对它们进行检查与调整；另一方面，需求分析又是软件设计、实现、测试直至维护的主要基础。良好的分析活动有助于避免或尽早剔除错误，从而提高软件生产率，降低开发成本，改进软件质量。

扫码观看本节视频讲解

11.2.1 可行性分析

考勤管理是企业管理中非常重要的一环，公司主管考勤的人员能够通过考勤管理系统清楚地看到公司员工的签到时间、签离时间以及是否迟到、早退等诸多信息，还能够通过所有员工的出勤记录比较来发现企业管理和员工作业方面的诸多问题，更是员工工资及福利待遇方面的重要参考依据。

11.2.2 系统操作流程分析

- 职工用户登录系统，上下班时进行签到考勤，经过系统验证通过后该员工签到成功。

- 管理用户登录本系统，输入用户名和密码，系统进行验证，验证通过的话进入程序主界面，在主界面对普通用户的信息进行录入，使用摄像头采集员工的人脸，然后通过机器学习技术创建学习模型。

11.2.3 系统模块设计

(1) 登录验证模块。

通过登录表单登录系统，整个系统分为管理员用户和普通用户。

(2) 考勤打卡。

普通用户登录系统后，可以分别实现在线上班打卡签到和下班打卡功能。

(3) 添加新用户信息。

管理员用户可以在后台添加新的员工信息，包括添加新员工的用户名和密码。

(4) 采集照片。

管理员用户可以在后台采集员工的照片，输入用户名，然后使用摄像头采集员工的照片。

(5) 训练照片模型。

使用机器学习技术训练采集到的员工照片，供员工打卡签到使用。

(6) 考勤统计管理。

使用可视化工具绘制员工的考勤数据，使用折线图统计最近两周的每天到场的员工人数。

本项目的功能模块如图 11-1 所示。

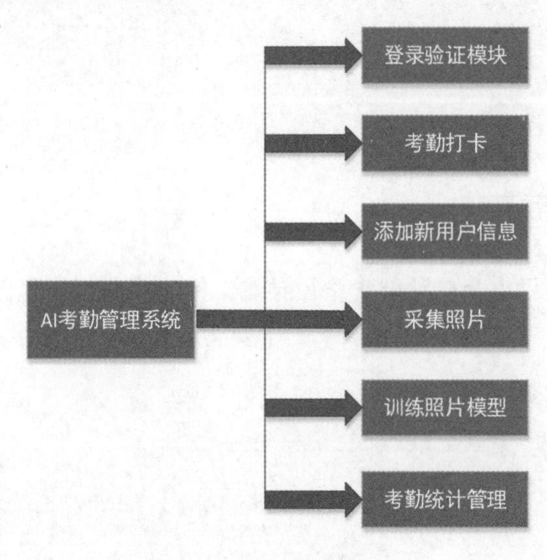

图 11-1 功能模块

11.3 系统配置

本系统是使用库 Django 实现的 Web 项目，在创建 Django Web 后会自动生成配置文件，开发者需要根据项目的需求设置这些配置文件。

扫码观看本节视频讲解

11.3.1 Django 配置文件

文件 settings.py 是 Django 项目的配置文件，主要用于设置整个 Django 项目所用到的程序文件和配置信息。在本项目中，需要设置 SQLite3 数据库的名字 db.sqlite3，并分别设置系统主页、登录页面和登录成功页面的 URL。文件 settings.py 的主要实现代码如下所示：

```
DATABASES = {
    'default': {
        'ENGINE': 'django.db.backends.sqlite3',
        'NAME': os.path.join(BASE_DIR, 'db.sqlite3'),
    }
}

STATIC_URL = '/static/'
CRISPY_TEMPLATE_PACK = 'bootstrap4'
LOGIN_URL='login'
LOGOUT_REDIRECT_URL = 'home'

LOGIN_REDIRECT_URL='dashboard'
```

11.3.2 路径导航文件

在 Django Web 项目中会自动创建路径导航文件 urls.py，在里面设置整个 Web 中所有页面对应的视图模块。本实例文件 urls.py 的主要实现代码如下所示：

```
urlpatterns = [
    path('admin/', admin.site.urls),
    path('', recog_views.home, name='home'),

    path('dashboard/', recog_views.dashboard, name='dashboard'),
    path('train/', recog_views.train, name='train'),
    path('add_photos/', recog_views.add_photos, name='add-photos'),
```

```
    path('login/',auth_views.LoginView.as_view(template_name=
        'users/login.html'),name='login'),
    path('logout/',auth_views.LogoutView.as_view(template_name=
        'recognition/home.html'),name='logout'),
    path('register/', users_views.register, name='register'),
    path('mark_your_attendance', recog_views.mark_your_attendance,
        name='mark-your-attendance'),
    path('mark_your_attendance_out', recog_views.mark_your_attendance_out,
        name='mark-your-attendance-out'),
    path('view_attendance_home', recog_views.view_attendance_home,
        name='view-attendance-home'),

    path('view_attendance_date', recog_views.view_attendance_date,
        name='view-attendance-date'),
    path('view_attendance_employee', recog_views.view_attendance_employee,
        name='view-attendance-employee'),
    path('view_my_attendance', recog_views.view_my_attendance_employee_login,
        name='view-my-attendance-employee-login'),
    path('not_authorised', recog_views.not_authorised, name='not-authorised')
]
```

11.4 用户注册和登录验证

为了提高开发效率，本项目使用库 Django 中的 django.contrib.auth 模块实现用户注册和登录验证功能。这样做的好处是减少代码编写量，节省开发时间。

扫码观看本节视频讲解

11.4.1 登录验证

根据文件 urls.py 中的如下代码可知，用户登录页面对应的模板文件是 login.html，此文件提供了用户登录表单，调用 django.contrib.auth 模块验证表单中的数据是否合法：

```
path('login/',auth_views.LoginView.as_view(template_name='users/login.html'),
name='login'),
```

文件 login.html 的主要实现代码如下所示：

```
{% load static %}
{% load crispy_forms_tags %}

<!DOCTYPE html>
```

```
<html>
<head>

   <!-- Bootstrap CSS -->
   <link rel="stylesheet" href="https://maxcdn.bootstrapcdn.com/bootstrap/
            4.0.0/css/bootstrap.min.css" integrity="sha384-Gn5384xqQ1aoWXA
            +058RXPxPg6fy4IWvTNh0E263XmFcJlSAwiGgFAW/dAiS6JXm" crossorigin
             ="anonymous">

  <style>
   body{
     background: url('{% static "recognition/img/bg_image.png"%}') no-repeat
                        center center fixed;
     background-size: cover;

   }

  </style>

</head>
<body>

 <div class="col-lg-12" style="background: rgba(0,0,0,0.6);max-height: 20px ;
padding-top:1em;padding-bottom:3em;color:#fff;border-radius:10px;
-webkit-box-shadow: 2px 2px 15px 0px rgba(0, 3, 0, 0.7);
-moz-box-shadow:   2px 2px 15px 0px rgba(0, 3, 0, 0.7);
box-shadow:       2px 2px 15px 0px rgba(0, 3, 0, 0.7); margin-left:auto;
margin-right: auto; ">

  <a href="{% url 'home' %}"><h5 class="text-left"> Home</h5></a>
</div>

 <div class="col-lg-4" style="background: rgba(0,0,0,0.6);margin-top:300px ;
padding-top:1em;padding-bottom:3em;color:#fff;border-radius:10px;
-webkit-box-shadow: 2px 2px 15px 0px rgba(0, 3, 0, 0.7);
-moz-box-shadow:   2px 2px 15px 0px rgba(0, 3, 0, 0.7);
box-shadow:       2px 2px 15px 0px rgba(0, 3, 0, 0.7); margin-left:auto;
margin-right: auto; ">

  <form method="POST" >
     {% csrf_token %}
     <fieldset class="form-group">
       <legend class="border-bottom mb-4"> Log In </legend>
       {{form| crispy}}
     </fieldset>

     <div class="form-group">
       <button class="btn btn-outline-info" type="submit"> Login!</button>
     </div>
```

```
    </form>

</div>
```

用户登录验证表单页面的执行效果如图 11-2 所示。

图 11-2 用户登录验证表单页面

11.4.2 添加新用户

根据文件 urls.py 中的如下代码可知，新用户注册页面对应的功能模块是 users_views.register：

```
path('register/', users_views.register, name='register'),
```

在文件 views.py 中，函数 register()用于获取注册表单中的注册信息，实现新用户注册功能。文件 views.py 的主要实现代码如下所示：

```
@login_required
def register(request):
    if request.user.username!='admin':
        return redirect('not-authorised')
    if request.method=='POST':
        form=UserCreationForm(request.POST)
        if form.is_valid():
            form.save() ###add user to database
            messages.success(request, f'Employee registered successfully!')
            return redirect('dashboard')

    else:
        form=UserCreationForm()
    return render(request,'users/register.html', {'form' : form})
```

在模板文件 register.html 中提供了注册表单功能，主要实现代码如下所示：

```
<form method="POST" >
    {% csrf_token %}
    <fieldset class="form-group">
      <legend class="border-bottom mb-4"> Register New Employee </legend>
      {{form| crispy}}
    </fieldset>
    <div class="form-group">
      <button class="btn btn-outline-info" type="submit"> Register</button>
    </div>
  </form>
</div>
```

添加新用户表单页面的执行效果如图 11-3 所示。

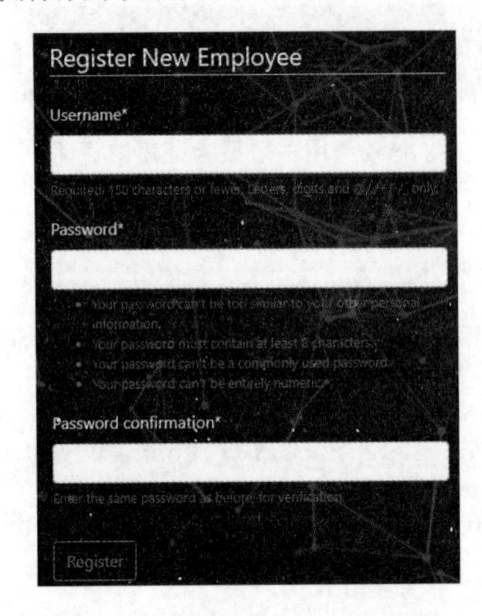

图 11-3　添加新用户表单页面

11.4.3　设计数据模型

在 Django Web 项目中，使用模型文件 models.py 设计项目中需要的数据库结构。因为本项目使用 django.contrib.auth 模块实现登录验证功能，所以在文件 models.py 中无须为会员用户设计数据库结构。模型文件 models.py 的主要实现代码如下所示：

```
from django.db import models
from django.contrib.auth.models import User

import datetime

class Present(models.Model):
    user=models.ForeignKey(User,on_delete=models.CASCADE)
```

```
    date = models.DateField(default=datetime.date.today)
    present=models.BooleanField(default=False)

class Time(models.Model):
    user=models.ForeignKey(User,on_delete=models.CASCADE)
    date = models.DateField(default=datetime.date.today)
    time=models.DateTimeField(null=True,blank=True)
    out=models.BooleanField(default=False)
```

通过上述代码设计了两个数据库表。

- Present：保存当前的打卡信息。
- Time：保存打卡的时间信息。

11.5 采集照片和机器学习

添加新的注册员工信息后，接下来需要采集员工的照片，然后使用 Scikit-Learn 将这些照片训练为机器学习模型，为员工的考勤打卡提供人脸识别和检测功能。

扫码观看本节视频讲解

11.5.1 设置采集对象

管理员用户成功登录系统后，来到后台主页 http://127.0.0.1:8000/dashboard/，执行效果如图 11-4 所示。

图 11-4　后台主页

管理员可以在后台采集员工的照片，单击 Add Photos 上面的“+”按钮后来到 http://127.0.0.1:8000/add_photos/，在此页面提供了如图 11-5 所示的表单，在表单中输入被采集对象的用户名。

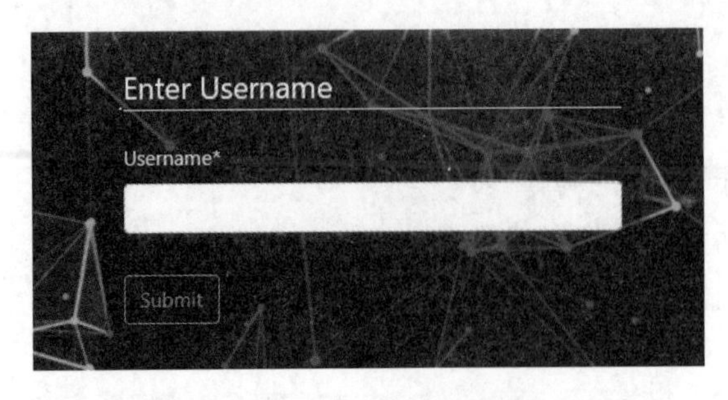

图 11-5　输入被采集对象的用户名

根据文件 urls.py 中的如下代码可知，输入被采集对象用户名页面对应的视图模块是 recog_views.add_photos：

```
    path('add_photos/', recog_views.add_photos, name='add-photos'),
```

在文件 views.py 中，视图函数 add_photos()的功能是获取在表单中输入的用户名，验证输入的用户名是否在数据库中存在，如果存在则继续下一步的照片采集工作。函数 add_photos()的具体实现代码如下所示：

```
@login_required
def add_photos(request):
    if request.user.username!='admin':
        return redirect('not-authorised')
    if request.method=='POST':
        form=usernameForm(request.POST)
        data = request.POST.copy()
        username=data.get('username')
        if username_present(username):
            create_dataset(username)
            messages.success(request, f'Dataset Created')
            return redirect('add-photos')
        else:
            messages.warning(request, f'No such username found. Please
                        register employee first.')
            return redirect('dashboard')
    else:


            form=usernameForm()
            return render(request,'recognition/add_photos.html', {'form' : form})
```

文件 add_photos.html 提供了输入被采集对象用户名的表单，主要实现代码如下所示：

```
<form method="POST" >
    {% csrf_token %}
    <fieldset class="form-group">
      <legend class="border-bottom mb-4"> Enter Username </legend>
      {{form| crispy}}
    </fieldset>

    <div class="form-group">
      <button class="btn btn-outline-info" type="submit"> Submit</button>
    </div>
   </form>
</div>
<div class="col-lg-12" style="padding-top: 100px;">
 {% if messages %}
    {% for message in messages%}
    <div class="alert alert-{{message.tags}}" > {{message}}
    </div>
    {%endfor %}

   {%endif%}
  </div>
```

11.5.2 采集照片

在采集表单中输入用户名并单击 Submit 按钮后，打开当前电脑中的摄像头采集照片，然后采集照片中的人脸，并将这些人脸数据创建为 Dataset 文件。在文件 views.py 中，视图函数 create_dataset()的功能是将采集的照片创建为 Dataset 文件，具体实现代码如下所示：

```
def create_dataset(username):
    id = username
    if(os.path.exists('face_recognition_data/training_dataset/{}/'.
       format(id))==False):

    os.makedirs('face_recognition_data/training_dataset/{}/'.format(id))
    directory='face_recognition_data/training_dataset/{}/'.format(id)

    #检测人脸
    print("[INFO] Loading the facial detector")
    detector = dlib.get_frontal_face_detector()
    predictor = dlib.shape_predictor('face_recognition_data/
                                     shape_predictor_68_face_landmarks.dat')
                                     #向形状预测器添加路径，稍后更改为相对路径
    fa = FaceAligner(predictor , desiredFaceWidth = 96)
```

```
# 从摄像头捕获图像并处理和检测人脸
# 初始化视频流
print("[INFO] Initializing Video stream")
vs = VideoStream(src=0).start()
#time.sleep(2.0) ####CHECK######

#识别码，将把 id 放在这里，并将 id 与一张脸一起存储，以便稍后可以识别它是谁的脸，
#设置初始值是 0
sampleNum = 0
#一张一张地捕捉人脸，检测出人脸并显示在窗口上
while(True):
    #拍摄图像，使用 vs.read 读取每一帧
    frame = vs.read()
    #调整每个图像的大小
    frame = imutils.resize(frame ,width = 800)
    #返回的 img 是一个彩色图像，但是为了使分类器工作，需要一个灰度图像来转换
    gray_frame = cv2.cvtColor(frame, cv2.COLOR_BGR2GRAY)
    #存储人脸，检测当前帧中的所有图像，并返回图像中人脸的坐标和其他一些参数以获得
    #准确的结果
    faces = detector(gray_frame,0)
    #在上面的 faces 变量中，可以有多个人脸，因此我们必须得到每个人脸，并在其周围绘制
    #一个矩形
    for face in faces:
        print("inside for loop")
        (x,y,w,h) = face_utils.rect_to_bb(face)

        face_aligned = fa.align(frame,gray_frame,face)
        #每当程序捕捉到人脸时，我们都会把它写成一个文件夹
        #在捕获人脸之前，我们需要告诉脚本它是为谁的人脸创建的，需要一个标识符，这里
        #称之为 id
        #抓到一张人脸后需要把它写进一个文件
        sampleNum = sampleNum+1
        #保存图像数据集，但只保存面部，裁剪掉其余的部分
        if face is None:
            print("face is none")
            continue

        cv2.imwrite(directory+'/'+str(sampleNum)+'.jpg' , face_aligned)
        face_aligned = imutils.resize(face_aligned ,width = 400)
        # cv2.imshow("Image Captured",face_aligned)
        # 下面对函数 rectangle()的参数的说明
        # @params 矩形的初始点是 x, y，终点是 x 的宽度和 y 的高度
        # @params 矩形的颜色
        # @params 矩形的厚度
        cv2.rectangle(frame,(x,y),(x+w,y+h),(0,255,0),1)
        # 在继续下一个循环之前，设置 50 毫秒的暂停等待键
```

```
        cv2.waitKey(50)

      #在另一个窗口中显示图像，创建一个窗口 Face，图像为 img
      cv2.imshow("Add Images",frame)
      #在关闭它之前，需要给出一个 wait 命令，否则 opencv 将无法工作，通过以下代码设置
      #延迟 1 毫秒
      cv2.waitKey(1)
      #跳出循环
      if(sampleNum>300):
          break

   #Stoping the videostream
   vs.stop()
   #销毁所有窗口
   cv2.destroyAllWindows()
```

11.5.3 训练照片模型

在创建 Dataset 文件后，单击后台主页中的 Train 图表按钮，使用机器学习技术 Scikit-Learn 训练 Dataset 数据集文件。根据文件 urls.py 中的如下代码可知，本项目通过 recog_views.train 模块训练 Dataset 数据集文件：

```
path('train/', recog_views.train, name='train'),
```

在视图文件 views.py 中，函数 predict()的功能是实现预测处理，具体实现代码如下所示：

```
def predict(face_aligned,svc,threshold=0.7):
    face_encodings=np.zeros((1,128))
    try:
        x_face_locations=face_recognition.face_locations(face_aligned)
        faces_encodings=face_recognition.face_encodings(face_aligned,
                        known_face_locations=x_face_locations)
        if(len(faces_encodings)==0):
            return ([-1],[0])
    except:
        return ([-1],[0])

    prob=svc.predict_proba(faces_encodings)
    result=np.where(prob[0]==np.amax(prob[0]))
    if(prob[0][result[0]]<=threshold):
        return ([-1],prob[0][result[0]])

    return (result[0],prob[0][result[0]])
```

在视图文件 views.py 中，函数 train()的功能是训练数据集文件，具体实现代码如下所示：

```
@login_required
def train(request):
    if request.user.username!='admin':
        return redirect('not-authorised')
    training_dir='face_recognition_data/training_dataset'

    count=0
    for person_name in os.listdir(training_dir):
        curr_directory=os.path.join(training_dir,person_name)
        if not os.path.isdir(curr_directory):
            continue
        for imagefile in image_files_in_folder(curr_directory):
            count+=1

    X=[]
    y=[]
    i=0

    for person_name in os.listdir(training_dir):
        print(str(person_name))
        curr_directory=os.path.join(training_dir,person_name)
        if not os.path.isdir(curr_directory):
            continue
        for imagefile in image_files_in_folder(curr_directory):
            print(str(imagefile))
            image=cv2.imread(imagefile)
            try:

    X.append((face_recognition.face_encodings(image)[0]).tolist())

                y.append(person_name)
                i+=1
            except:
                print("removed")
                os.remove(imagefile)

    targets=np.array(y)
    encoder = LabelEncoder()
    encoder.fit(y)
    y=encoder.transform(y)
    X1=np.array(X)
    print("shape: "+ str(X1.shape))
    np.save('face_recognition_data/classes.npy', encoder.classes_)
    svc = SVC(kernel='linear',probability=True)
    svc.fit(X1,y)
    svc_save_path="face_recognition_data/svc.sav"
    with open(svc_save_path, 'wb') as f:
```

```
        pickle.dump(svc,f)
    vizualize_Data(X1,targets)
    messages.success(request, f'Training Complete.')
    return render(request,"recognition/train.html")
```

训练完毕后会可视化展示训练结果，如图 11-6 所示，说明本项目在目前只是采集了两名员工的照片信息。

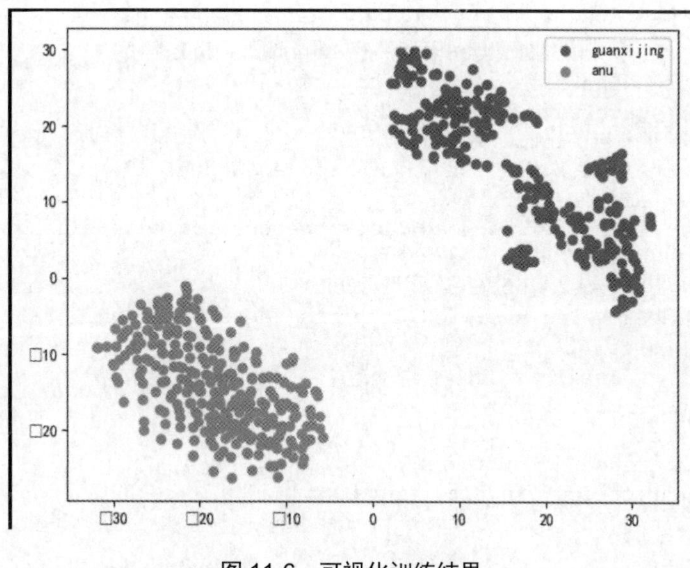

图 11-6 可视化训练结果

11.6 考勤打卡

员工登录系统主页后，可以分别实现在线上班打卡签到和下班打卡功能。在本节的内容中，将详细讲解实现考勤打卡功能的过程。

扫码观看本节视频讲解

11.6.1 上班打卡签到

在系统主页单击 Mark Your Attendance - In 上面的图标链接，来到上班打卡页面 http://127.0.0.1:8000/mark_your_attendance，根据文件 urls.py 中的如下代码可知，考勤打卡页面功能是通过调用 recog_views.mark_your_attendance 模块实现的：

```
path('mark_your_attendance', recog_views.mark_your_attendance,
name='mark-your-attendance')
```

在视图文件 views.py 中，函数 mark_your_attendance(request)的功能是采集摄像头中的人脸，根据前面训练的模型识别出是哪一名员工，然后实现考勤打卡功能，并将打卡信息添加到数据库中。函数 mark_your_attendance(request)的具体实现代码如下所示：

```
def mark_your_attendance(request):
    detector = dlib.get_frontal_face_detector()
    predictor = dlib.shape_predictor('face_recognition_data/
                shape_predictor_68_face_landmarks.dat')
                #向形状预测器中添加路径，稍后更改为相对路径
    svc_save_path="face_recognition_data/svc.sav"
    with open(svc_save_path, 'rb') as f:
            svc = pickle.load(f)
    fa = FaceAligner(predictor, desiredFaceWidth = 96)
    encoder=LabelEncoder()
    encoder.classes_ = np.load('face_recognition_data/classes.npy')

    faces_encodings = np.zeros((1,128))
    no_of_faces = len(svc.predict_proba(faces_encodings)[0])
    count = dict()
    present = dict()
    log_time = dict()
    start = dict()
    for i in range(no_of_faces):
        count[encoder.inverse_transform([i])[0]] = 0
        present[encoder.inverse_transform([i])[0]] = False

    vs = VideoStream(src=0).start()
    sampleNum = 0
    while(True):
        frame = vs.read()
        frame = imutils.resize(frame ,width = 800)
        gray_frame = cv2.cvtColor(frame, cv2.COLOR_BGR2GRAY)
        faces = detector(gray_frame,0)

        for face in faces:
            print("INFO : inside for loop")
            (x,y,w,h) = face_utils.rect_to_bb(face)
            face_aligned = fa.align(frame,gray_frame,face)
            cv2.rectangle(frame,(x,y),(x+w,y+h),(0,255,0),1)
            (pred,prob)=predict(face_aligned,svc)
            if(pred!=[-1]):

                person_name=encoder.inverse_transform(np.ravel([pred]))[0]
                pred=person_name
                if count[pred] == 0:
                    start[pred] = time.time()
```

```
                    count[pred] = count.get(pred,0) + 1

                if count[pred] == 4 and (time.time()-start[pred]) > 1.2:
                     count[pred] = 0
                else:
                #if count[pred] == 4 and (time.time()-start) <= 1.5:
                    present[pred] = True
                    log_time[pred] = datetime.datetime.now()
                    count[pred] = count.get(pred,0) + 1
                    print(pred, present[pred], count[pred])
                cv2.putText(frame, str(person_name)+ str(prob), (x+6,y+h-6),
                                 cv2.FONT_HERSHEY_SIMPLEX,0.5,(0,255,0),1)
            else:
                person_name="unknown"
                cv2.putText(frame, str(person_name), (x+6,y+h-6),
                            cv2.FONT_HERSHEY_SIMPLEX,0.5,(0,255,0),1)

            #cv2.putText()
            #在继续下一个循环之前，设置一个 50 毫秒的暂停等待键
            #cv2.waitKey(50)

        #在另一个窗口中显示图像
        #创建一个窗口 Face，图像为 img
        cv2.imshow("Mark Attendance - In - Press q to exit",frame)
        #在关闭它之前，我们需要给出一个 wait 命令，否则 opencv 将无法工作
        #下面的参数 1 表示延迟 1 毫秒
        #cv2.waitKey(1)
        #停止循环
        key=cv2.waitKey(50) & 0xFF
        if(key==ord("q")):
            break
    #停止视频流
    vs.stop()

    #销毁所有窗体
    cv2.destroyAllWindows()
    update_attendance_in_db_in(present)
    return redirect('home')
```

11.6.2 下班打卡

在系统主页单击 Mark Your Attendance - Out 上面的图标链接，来到下班打卡页面 http://127.0.0.1:8000/mark_your_attendance_out，根据文件 urls.py 中的如下代码可知，下班打卡页面功能是通过调用 recog_views.mark_your_attendance_out 模块实现的：

```
path('mark_your_attendance_out', recog_views.mark_your_attendance_out,
                     name='mark-your-attendance-out')
```

在视图文件 views.py 中，函数 mark_your_attendance_out()的功能是采集摄像头中的人脸，根据前面训练的模型识别出是哪一名员工，然后实现下班打卡功能，并将打卡信息添加到数据库中。函数 mark_your_attendance_out()的具体实现代码如下所示：

```
def mark_your_attendance_out(request):
    detector = dlib.get_frontal_face_detector()
    predictor = dlib.shape_predictor('face_recognition_data/
                     shape_predictor_68_face_landmarks.dat')
                     #向形状预测器添加路径，稍后更改为相对路径
    svc_save_path="face_recognition_data/svc.sav"

    with open(svc_save_path, 'rb') as f:
            svc = pickle.load(f)
    fa = FaceAligner(predictor , desiredFaceWidth = 96)
    encoder=LabelEncoder()
    encoder.classes_ = np.load('face_recognition_data/classes.npy')

    faces_encodings = np.zeros((1,128))
    no_of_faces = len(svc.predict_proba(faces_encodings)[0])
    count = dict()
    present = dict()
    log_time = dict()
    start = dict()
    for i in range(no_of_faces):
        count[encoder.inverse_transform([i])[0]] = 0
        present[encoder.inverse_transform([i])[0]] = False

    vs = VideoStream(src=0).start()
    sampleNum = 0
    while(True):
        frame = vs.read()
        frame = imutils.resize(frame ,width = 800)
        gray_frame = cv2.cvtColor(frame, cv2.COLOR_BGR2GRAY)
        faces = detector(gray_frame,0)

        for face in faces:
            print("INFO : inside for loop")
            (x,y,w,h) = face_utils.rect_to_bb(face)
            face_aligned = fa.align(frame,gray_frame,face)
            cv2.rectangle(frame,(x,y),(x+w,y+h),(0,255,0),1)

            (pred,prob)=predict(face_aligned,svc)
            if(pred!=[-1]):
```

```
                person_name=encoder.inverse_transform(np.ravel([pred]))[0]
                pred=person_name
                if count[pred] == 0:
                    start[pred] = time.time()
                    count[pred] = count.get(pred,0) + 1
                if count[pred] == 4 and (time.time()-start[pred]) > 1.5:
                     count[pred] = 0
                else:
                #if count[pred] == 4 and (time.time()-start) <= 1.5:
                    present[pred] = True
                    log_time[pred] = datetime.datetime.now()
                    count[pred] = count.get(pred,0) + 1
                    print(pred, present[pred], count[pred])
                cv2.putText(frame, str(person_name)+ str(prob), (x+6,y+h-6),
                            cv2.FONT_HERSHEY_SIMPLEX,0.5,(0,255,0),1)
            else:
                person_name="unknown"
                cv2.putText(frame, str(person_name), (x+6,y+h-6),
                            cv2.FONT_HERSHEY_SIMPLEX,0.5,(0,255,0),1)


        #在另一个窗口中显示图像将创建一个窗口 Face，图像为 img
        cv2.imshow("Mark Attendance- Out - Press q to exit",frame)
        #在关闭它之前，我们需要给出一个 wait 命令，否则 opencv 将无法工作，
        #下面的参数@params 表示延迟 1 毫秒
        # cv2.waitKey(1)
        key=cv2.waitKey(50) & 0xFF
        if(key==ord("q")):
            break
    vs.stop()

    cv2.destroyAllWindows()
    update_attendance_in_db_out(present)
    return redirect('home')
```

11.7 可视化考勤数据

管理登录系统后，可以在考勤统计管理页面查看员工的考勤信息。在本项目中，使用可视化工具绘制员工的考勤数据，使用折线图统计最近两周的员工考勤信息。

扫码观看本节视频讲解

11.7.1 统计最近两周的考勤数据

1. 视图函数

在后台主页单击 View Attendance Reports 上面的图标链接，在打开的网页 http://127.0.0.1:8000/view_attendance_home 中可以查看员工的考勤统计信息。根据文件 urls.py 中的如下代码可知，可视化考勤数据页面的功能是通过调用 recog_views.view_attendance_home 模块实现的：

```
path('view_attendance_home', recog_views.view_attendance_home,
                             name='view-attendance-home')
```

在视图文件 views.py 中，函数 view_attendance_home()的功能是可视化展示员工的考勤信息，具体实现代码如下所示：

```
@login_required
def view_attendance_home(request):
    total_num_of_emp=total_number_employees()
    emp_present_today=employees_present_today()
    this_week_emp_count_vs_date()
    last_week_emp_count_vs_date()
    return render(request,"recognition/view_attendance_home.html",
{'total_num_of_emp' : total_num_of_emp, 'emp_present_today':
emp_present_today})
```

在上述代码中用到了如下所示的 4 个函数。

(1) 函数 total_number_employees()的功能是统计当前系统中的考勤员工信息，具体实现代码如下所示：

```
def total_number_employees():
    qs=User.objects.all()
    return (len(qs) -1)
```

(2) 函数 employees_present_today()的功能是统计今日打卡的员工数量，具体实现代码如下所示：

```
def employees_present_today():
    today=datetime.date.today()
    qs=Present.objects.filter(date=today).filter(present=True)
    return len(qs)
```

(3) 函数 this_week_emp_count_vs_date()的功能是统计本周每天员工的打卡信息，并绘制可视化折线图。具体实现代码如下所示：

```
def this_week_emp_count_vs_date():
    today=datetime.date.today()
    some_day_last_week=today-datetime.timedelta(days=7)
    monday_of_last_week=some_day_last_week-
datetime.timedelta(days=(some_day_last_week.isocalendar()[2] - 1))
    monday_of_this_week = monday_of_last_week + datetime.timedelta(days=7)
    qs=Present.objects.filter(date__gte=monday_of_this_week).
                                filter(date__lte=today)
    str_dates=[]
    emp_count=[]
    str_dates_all=[]
    emp_cnt_all=[]
    cnt=0

    for obj in qs:
        date=obj.date
        str_dates.append(str(date))
        qs=Present.objects.filter(date=date).filter(present=True)
        emp_count.append(len(qs))
    while(cnt<5):
        date=str(monday_of_this_week+datetime.timedelta(days=cnt))
        cnt+=1
        str_dates_all.append(date)
        if(str_dates.count(date))>0:
            idx=str_dates.index(date)
            emp_cnt_all.append(emp_count[idx])
        else:
            emp_cnt_all.append(0)

    df=pd.DataFrame()
    df["date"]=str_dates_all
    df["Number of employees"]=emp_cnt_all

    sns.lineplot(data=df,x='date',y='Number of employees')
    plt.savefig('./recognition/static/recognition/img/attendance_graphs/
                this_week/1.png')
    plt.close()
```

(4) 函数 last_week_emp_count_vs_date()的功能是统计上一周每天员工的打卡信息，具体实现代码如下所示：

```
def last_week_emp_count_vs_date():
    today=datetime.date.today()
    some_day_last_week=today-datetime.timedelta(days=7)
    monday_of_last_week=some_day_last_week-  datetime.timedelta
                        (days=(some_day_last_week.isocalendar()[2] - 1))
    monday_of_this_week = monday_of_last_week + datetime.timedelta(days=7)
```

```
qs=Present.objects.filter(date__gte=monday_of_last_week).
                              filter(date__lt=monday_of_this_week)
str_dates=[]
emp_count=[]
str_dates_all=[]
emp_cnt_all=[]
cnt=0

for obj in qs:
    date=obj.date
    str_dates.append(str(date))
    qs=Present.objects.filter(date=date).filter(present=True)
    emp_count.append(len(qs))
while(cnt<5):
    date=str(monday_of_last_week+datetime.timedelta(days=cnt))
    cnt+=1
    str_dates_all.append(date)
    if(str_dates.count(date))>0:
        idx=str_dates.index(date)
        emp_cnt_all.append(emp_count[idx])
    else:
        emp_cnt_all.append(0)
df=pd.DataFrame()
df["date"]=str_dates_all
df["emp_count"]=emp_cnt_all

sns.lineplot(data=df,x='date',y='emp_count')
plt.savefig('./recognition/static/recognition/img/attendance_graphs/
             last_week/1.png')
plt.close()
```

2. 模板文件

编写模板文件 view_attendance_home.html，功能是调用上面的视图函数，使用曲线图可视化展示最近两周的员工考勤数据。主要实现代码如下所示：

```
<div class="collapse navbar-collapse" id="navbarNav">
  <ul class="navbar-nav">

    <li class="nav-item active">
      <a class="nav-link" href="{%url 'view-attendance-employee' %}">
             By Employee</a>
    </li>
      <li class="nav-item active">
      <a class="nav-link" href="{% url 'view-attendance-date' %}">By Date</a>
    </li>
```

```
        <li class="nav-item active" style="padding-left: 1440px">
       <a class="nav-link" href="{% url 'dashboard' %}">Back to Admin Panel</a>
     </li>
   </ul>
  </div>
</nav>

  <div class="card" style="margin-top: 2em; margin-left: 2em; margin-right: 2em;
                           margin-bottom: 2em;">
   <div class="card-body">
<h2> Today's Statistics </h2>
    <div class="row" style="margin-left: 12em">
<div class="card" style="width: 20em; background-color: #338044; text-align :
                    center; margin-left: 5em; margin-top: 5em; color: white;">
  <div class="card-body">
   <h5 class="card-title"> <b>Total Number Of Employees</b></h5>
   <p class="card-text" style="padding-top: 1em; font-size: 28px;">
<b>{{total_num_of_emp }}</b></p>
  </div>
</div>
<div class="card" style="width: 20em; background-color: #80335b; text-align :
                  center; margin-left: 5em; margin-top: 5em; color: white;">
  <div class="card-body">
   <h5 class="card-title"> <b> Employees present today</b></h5>
   <p class="card-text" style="padding-top: 1em; font-size: 28px;"><b>
{{emp_present_today }}</b></p>
  </div>
</div>

</div>
</div>
</div>

  <div class="card" style="margin-top: 2em; margin-left: 2em; margin-right: 2em;
                           margin-bottom: 2em;">
   <div class="card-body">
  <div class="row" >
<div class="col-md-6">
<h2> Last Week </h2>
   <div class="card" style="width: 50em;">
  <img class="card-img-top" src="{% static
'recognition/img/attendance_graphs/last_week/1.png'%}" alt="Card image cap">
  <div class="card-body">
   <p class="card-text" style="text-align: center;">Number of employees
            present each day</p>
  </div>
</div>
```

```
</div>
<div class="col-md-6">
  <h2> This Week </h2>
 <div class="card" style="width: 50em;">
  <img class="card-img-top" src="{% static
'recognition/img/attendance_graphs/this_week/1.png'%}" alt="Card image cap">
  <div class="card-body">
    <p class="card-text" style="text-align: center;">Number of employees
              present each day</p>
  </div>
</div>
```

员工考勤数据可视化页面的执行效果如图 11-7 所示。

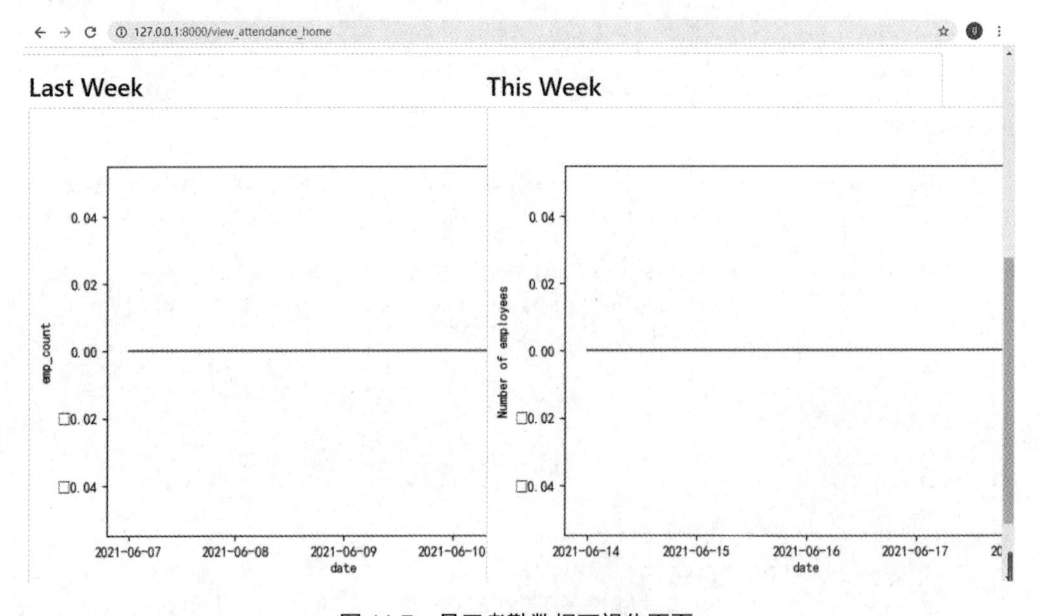

图 11-7　员工考勤数据可视化页面

11.7.2　查看本人指定时间范围内的考勤统计图

1. 视图函数

普通员工登录系统后，单击 View My Attendance 上面的图标链接，来到 http://127.0.0.1:8000/view_my_attendance 页面，如图 11-8 所示。

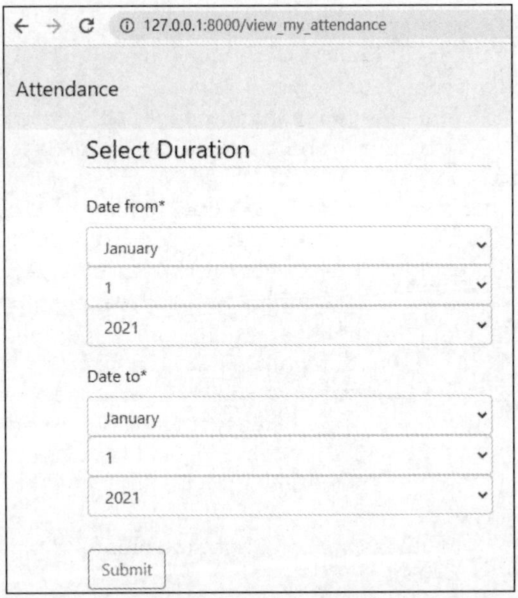

图 11-8 选择时间范围

在此页面中可以查看本人在指定时间段内的考勤统计图信息。根据文件 urls.py 中的如下代码可知，查看本人指定时间范围内的考勤数据的功能是通过调用 recog_views.view_my_attendance_employee_login 模块实现的：

```
path('view_my_attendance', recog_views.view_my_attendance_employee_login,
     name='view-my-attendance-employee-login')
```

在视图文件 views.py 中，函数 view_my_attendance_employee_login()的功能是可视化展示本人在指定时间段内的考勤信息，具体实现代码如下所示：

```
@login_required
def view_my_attendance_employee_login(request):
    if request.user.username=='admin':
        return redirect('not-authorised')
    qs=None
    time_qs=None
    present_qs=None
    if request.method=='POST':
        form=DateForm_2(request.POST)
        if form.is_valid():
            u=request.user
            time_qs=Time.objects.filter(user=u)
            present_qs=Present.objects.filter(user=u)
```

```
            date_from=form.cleaned_data.get('date_from')
            date_to=form.cleaned_data.get('date_to')
            if date_to < date_from:
                    messages.warning(request, f'Invalid date selection.')
                    return redirect('view-my-attendance-employee-login')
            else:
                    time_qs=time_qs.filter(date__gte=date_from).
                            filter(date__lte=date_to).order_by('-date')
                    present_qs=present_qs.filter(date__gte=date_from).
                            filter(date__lte=date_to).order_by('-date')

                    if (len(time_qs)>0 or len(present_qs)>0):
                        qs=hours_vs_date_given_employee
                                (present_qs,time_qs,admin=False)
                        return render(request,'recognition/view_my_attendance_
                            employee_login.html', {'form' : form, 'qs' :qs})
                    else:
                        messages.warning(request, f'No records for selected
                                          duration.')
                        return redirect('view-my-attendance-employee-login')
    else:
            form=DateForm_2()
            return render(request,'recognition/view_my_attendance_employee_
                        login.html', {'form' : form, 'qs' :qs})
```

在上述代码中，调用函数 hours_vs_date_given_employee()绘制在指定时间段内的考勤统计图，具体实现代码如下所示：

```
def hours_vs_date_given_employee(present_qs,time_qs,admin=True):
    register_matplotlib_converters()
    df_hours=[]
    df_break_hours=[]
    qs=present_qs
    for obj in qs:
        date=obj.date
        times_in=time_qs.filter(date=date).filter(out=False).order_by('time')
        times_out=time_qs.filter(date=date).filter(out=True).order_by('time')
        times_all=time_qs.filter(date=date).order_by('time')
        obj.time_in=None
        obj.time_out=None
        obj.hours=0
        obj.break_hours=0
        if (len(times_in)>0):
            obj.time_in=times_in.first().time
        if (len(times_out)>0):
```

```
            obj.time_out=times_out.last().time
        if(obj.time_in is not None and obj.time_out is not None):
            ti=obj.time_in
            to=obj.time_out
            hours=((to-ti).total_seconds())/3600
            obj.hours=hours
        else:
            obj.hours=0
        (check,break_hourss)= check_validity_times(times_all)
        if check:
            obj.break_hours=break_hourss

        else:
            obj.break_hours=0
        df_hours.append(obj.hours)
        df_break_hours.append(obj.break_hours)
        obj.hours=convert_hours_to_hours_mins(obj.hours)
        obj.break_hours=convert_hours_to_hours_mins(obj.break_hours)
    df = read_frame(qs)
    df["hours"]=df_hours
    df["break_hours"]=df_break_hours

    print(df)
    sns.barplot(data=df,x='date',y='hours')
    plt.xticks(rotation='vertical')
    rcParams.update({'figure.autolayout': True})
    plt.tight_layout()
    if(admin):
        plt.savefig('./recognition/static/recognition/img/attendance_
                    graphs/hours_vs_date/1.png')
        plt.close()
    else:
        plt.savefig('./recognition/static/recognition/img/attendance_
                    graphs/employee_login/1.png')
        plt.close()
    return qs
```

在上述代码中，如果当前登录用户是管理员，则绘制在指定时间段内本人每天的上班时间。如果当前登录用户不是管理员，而是普通员工，则绘制本人在这个时间段内的考勤统计图，如图 11-9 所示。

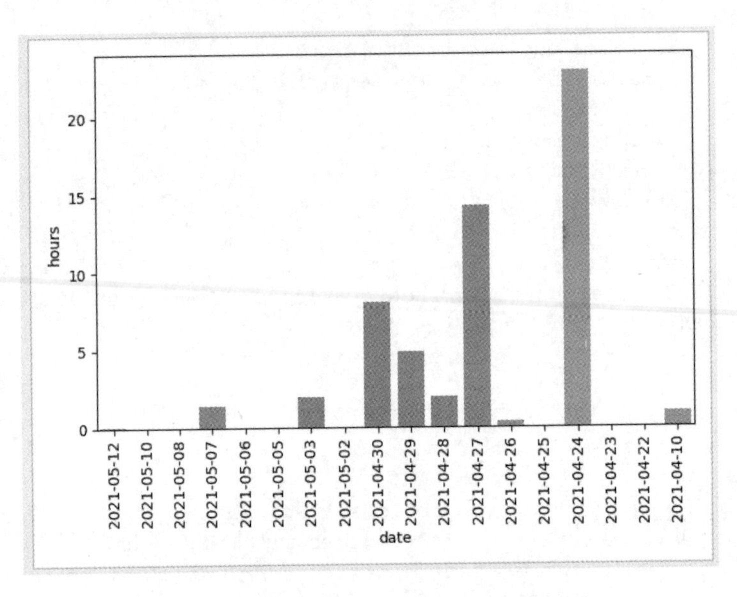

图 11-9 统计在指定时间段内的考勤信息

2. 模板文件

编写模板文件 view_attendance_date.html，功能是创建选择时间段的表单页面。主要实现代码如下所示：

```
<nav class="navbar navbar-expand-lg navbar-light bg-light">
<a class="navbar-brand" href="{%url 'view-attendance-home' %}">Attendance
          Dashboard</a>
<button class="navbar-toggler" type="button" data-toggle="collapse"
              data-target="#navbarNav" aria-controls="navbarNav"
              aria-expanded="false" aria-label="Toggle navigation">
  <span class="navbar-toggler-icon"></span>
</button>
<div class="collapse navbar-collapse" id="navbarNav">
  <ul class="navbar-nav">

    <li class="nav-item active">
      <a class="nav-link" href="{%url 'view-attendance-employee' %}">
              By Employee</a>
    </li>
      <li class="nav-item active">
      <a class="nav-link" href="{% url 'view-attendance-date' %}">By Date</a>
    </li>
       <li class="nav-item active" style="padding-left: 1440px">
      <a class="nav-link" href="{% url 'dashboard' %}">Back to Admin Panel</a>
    </li>
```

```
    </ul>
  </div>
</nav>

<div class="container">
  <div style="width: 400px">
 <form method="POST" >
      {% csrf_token %}
      <fieldset class="form-group">
        <legend class="border-bottom mb-4"> Select Date </legend>
        {{form| crispy}}
      </fieldset>
      <div class="form-group">
        <button class="btn btn-outline-info" type="submit" value="Create">
                         Submit</button>
      </div>
    </form>
</div>
{% if qs %}
<table  class="table" style="margin-top: 5em; ">
    <thead class="thead-dark">
    <tr>
        <th scope="col">Date</th>
        <th scope="col" >Employee</th>
        <th scope="col">Present</th>
        <th scope="col">Time in</th>
        <th scope="col">Time out </th>
        <th scope="col">Hours </th>
        <th scope="col"> Break Hours </th>
    </tr>
</thead>
<tbody>
    {% for item in qs %}
    <tr>
           <td>{{ item.date }}</td>
        <td>{{ item.user.username}}</td>
     {% if item.present %}
        <td> P </td>
        {% else %}
        <td> A </td>
        {% endif %}
        {% if item.time_in %}
        <td>{{ item.time_in }}</td>
      {% else %}
      <td> - </td>
      {% endif %}
```

```
        {% if item.time_out %}
       <td>{{ item.time_out }}</td>
      {% else %}
      <td> - </td>
      {% endif %}
         <td> {{item.hours}}</td>
         <td> {{item.break_hours}}</td>
   </tr>
   {% endfor %}
</tbody>
</table>
```

11.7.3 查看某员工在指定时间范围内的考勤统计图

1. 视图函数

管理员登录系统后，输入 URL 链接 http://127.0.0.1:8000/view_attendance_employee，如图 11-10 所示。在此页面中可以输入员工的名字和时间段，单击 Submit 按钮后可以查看这名员工在指定时间段内的考勤信息。

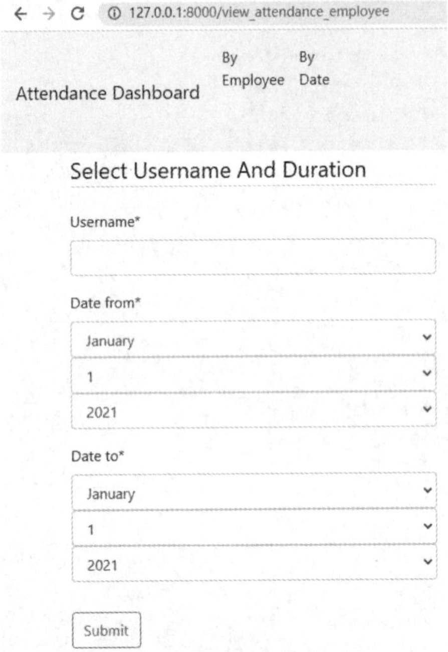

图 11-10 选择时间范围

在此页面中可以查看某员工在指定时间段内的考勤统计图信息。根据文件 urls.py 中的如下代码可知，查看某员工在指定时间范围内的考勤统计图的功能是通过调用 recog_views.view_attendance_employee 模块实现的：

```
path('view_attendance_employee', recog_views.view_attendance_employee,
                name='view-attendance-employee')
```

在视图文件 views.py 中，函数 view_attendance_employee()的功能是可视化展示指定员工在指定时间段内的考勤信息，具体实现代码如下所示：

```
@login_required
def view_attendance_employee(request):
    if request.user.username!='admin':
        return redirect('not-authorised')
    time_qs=None
    present_qs=None
    qs=None

    if request.method=='POST':
        form=UsernameAndDateForm(request.POST)
        if form.is_valid():
            username=form.cleaned_data.get('username')
            if username_present(username):
                u=User.objects.get(username=username)
                time_qs=Time.objects.filter(user=u)
                present_qs=Present.objects.filter(user=u)
                date_from=form.cleaned_data.get('date_from')
                date_to=form.cleaned_data.get('date_to')
                if date_to < date_from:
                    messages.warning(request, f'Invalid date selection.')
                    return redirect('view-attendance-employee')
                else:
                    time_qs=time_qs.filter(date__gte=date_from).
                            filter(date__lte=date_to).order_by('-date')
                    present_qs=present_qs.filter(date__gte=date_from).
                            filter(date__lte=date_to).order_by('-date')

                    if (len(time_qs)>0 or len(present_qs)>0):
                        qs=hours_vs_date_given_employee(present_qs,
                            time_qs,admin=True)
                        return render(request,'recognition/view_attendance_
                                employee.html', {'form' : form, 'qs' :qs})
                    else:
                        #print("inside qs is None")
                        messages.warning(request, f'No records for selected
                                        duration.')
```

```
                        return redirect('view-attendance-employee')

                else:
                    print("invalid username")
                    messages.warning(request, f'No such username found.')
                    return redirect('view-attendance-employee')
        else:
                form=UsernameAndDateForm()
               return render(request,'recognition/view_attendance_employee.html',
                        {'form' : form, 'qs' :qs})
```

在上述代码中，也需要使用前面介绍的视图函数 hours_vs_date_given_employee()绘制柱状考勤统计图。

2. 模板文件

编写模板文件 view_attendance_employee.html，功能是创建设置员工用户名和选择时间段的表单页面。主要实现代码如下所示：

```
<body>
    <nav class="navbar navbar-expand-lg navbar-light bg-light">
  <a class="navbar-brand" href="{%url 'view-attendance-home' %}">Attendance
          Dashboard</a>
  <button class="navbar-toggler" type="button" data-toggle="collapse"
              data-target="#navbarNav" aria-controls="navbarNav"
              aria-expanded="false" aria-label="Toggle navigation">
    <span class="navbar-toggler-icon"></span>
  </button>
  <div class="collapse navbar-collapse" id="navbarNav">
    <ul class="navbar-nav">

      <li class="nav-item active">
        <a class="nav-link" href="{%url 'view-attendance-employee' %}">
              By Employee</a>
      </li>
        <li class="nav-item active">
        <a class="nav-link" href="{% url 'view-attendance-date' %}">By Date</a>
      </li>
         <li class="nav-item active" style="padding-left: 1440px">
        <a class="nav-link" href="{% url 'dashboard' %}">Back to Admin Panel</a>
      </li>

    </ul>
  </div>
</nav>
```

```
<div class="container">
  <div style="width:400px;">

 <form method="POST" >
     {% csrf_token %}
     <fieldset class="form-group">
       <legend class="border-bottom mb-4"> Select Username And Duration </legend>
       {{form| crispy}}
     </fieldset>

     <div class="form-group">
       <button class="btn btn-outline-info" type="submit"> Submit</button>
     </div>
   </form>

</div>

{%if qs%}
<table class="table"  style="margin-top: 5em;">
   <thead class="thead-dark">
   <tr>
       <th scope="col">Date</th>

       <th scope="col" >Employee</th>
       <th scope="col">Present</th>
       <th scope="col">Time in</th>
       <th scope="col">Time out </th>
        <th scope="col">Hours </th>
         <th scope="col"> Break Hours </th>
   </tr>
</thead>
<tbody>
   {% for item in qs %}
   <tr>
         <td>{{ item.date }}</td>
       <td>{{ item.user.username}}</td>

        {% if item.present %}
       <td> P </td>
       {% else %}
       <td> A </td>
       {% endif %}
       {% if item.time_in %}
       <td>{{ item.time_in }}</td>
      {% else %}
      <td> - </td>
      {% endif %}
```

```
      {% if item.time_out %}
      <td>{{ item.time_out }}</td>
     {% else %}
     <td> - </td>
     {% endif %}
     <td> {{item.hours}}</td>
        <td> {{item.break_hours}}</td>
   </tr>
   {% endfor %}
</tbody>
</table>

 <div class="card" style=" margin-top: 5em; margin-bottom: 10em;">
  <img class="card-img-top" src="{% static 'recognition/img/attendance_graphs/
                              hours_vs_date/1.png'%}" alt="Card image cap">
  <div class="card-body">
    <p class="card-text" style="text-align: center;">Number of hours worked each
               day.</p>
  </div>
</div>
{% endif %}
 {% if messages %}
     {% for message in messages%}
     <div class="alert alert-{{message.tags}}" > {{message}}
     </div>
     {%endfor %}
   {%endif%}
```

第12章

AI小区停车计费管理系统

在国内经济飞速发展的今天，小区停车难、高峰期进出小区困难的问题日益严重。通过使用AI技术自动识别车牌，并实现实时计费功能可以提高车辆的通行率。在本章的内容中，将详细介绍使用百度AI技术开发一个小区停车计费管理系统的过程。

12.1 背景介绍

随着中国城镇化进程发展快速，目前城镇化水平已经超过中等收入国家平均水平的50%，接近高收入国家平均水平的70%，作为城镇的核心单位，住宅小区数量也在迅猛递增，目前全国的住宅小区数量已经超过30万个。小汽车作为居民的重要出行工具之一，目前全国汽车保有量已经高达2.6亿辆，小区停车难、高峰期进出小区困难的问题也持续了相当长的时间，一直得不到好的解决。

扫码观看本节视频讲解

除了停车难和停车场内拥堵，高昂的值班人员人工成本问题(每个车行通道需要2～3个值班收费人员)、收费跑冒滴漏的现象也屡禁不止。

信息技术的快速进步，让较低的投入实现降本增收成为现实。通过智能化成套系统方案，让小区通道实现无人值守化管理。针对既有临时车又有固定车的停车场，推出车牌识别停车计费管理系统方案，利用车牌识别技术取代传统的IC卡技术，解决了车辆进出时必须停下刷卡而造成的停车场进出口塞车现象，这种智能的停车管理系统为目前的停车场用户提供了一种崭新的服务模式。通过该系统，固定车辆进出可以实现不停车通行，临时车入场不停车、出场缴费自动放行，整个系统结构简单、稳定可靠，安装、维护、使用方便。

12.2 系统功能分析和模块设计

本小区停车计费管理系统是基于百度人工智能技术实现的，节省了物业的人员成本，提高了小区车辆的通行率，为大家的出行带来了极大的方便。

扫码观看本节视频讲解

12.2.1 功能分析

区别于传统的IC卡、蓝牙卡停车计费模式，本系统采用百度AI技术识别车牌信息，通过车牌号码判别是临时车还是固定车辆，从而对进出车辆进行计费管理。应用车牌识别停车计费系统具有以下特点。

(1) 固定车辆进场、出场都无须刷卡，车牌识别后自动放行，车辆进出效率明显提升。

(2) 临时车辆进场无须停车取卡，进场车牌识别后道闸自动放行，出场时车牌识别停车缴费后，道闸自动放行。

(3) 车牌识别停车管理模式免去了物业管理IC卡、蓝牙卡的成本，后续维护成本低。

(4) 不论是固定车还是临时车，车主都无须为卡未带、卡丢失、卡损坏而发愁，车主用户体验大大提升。

(5) 由于采用车牌识别停车管理模式后，车辆进出效率大大提高，物业停车收益也会明显提升。

12.2.2 系统模块设计

本项目的功能模块如图12-1所示。

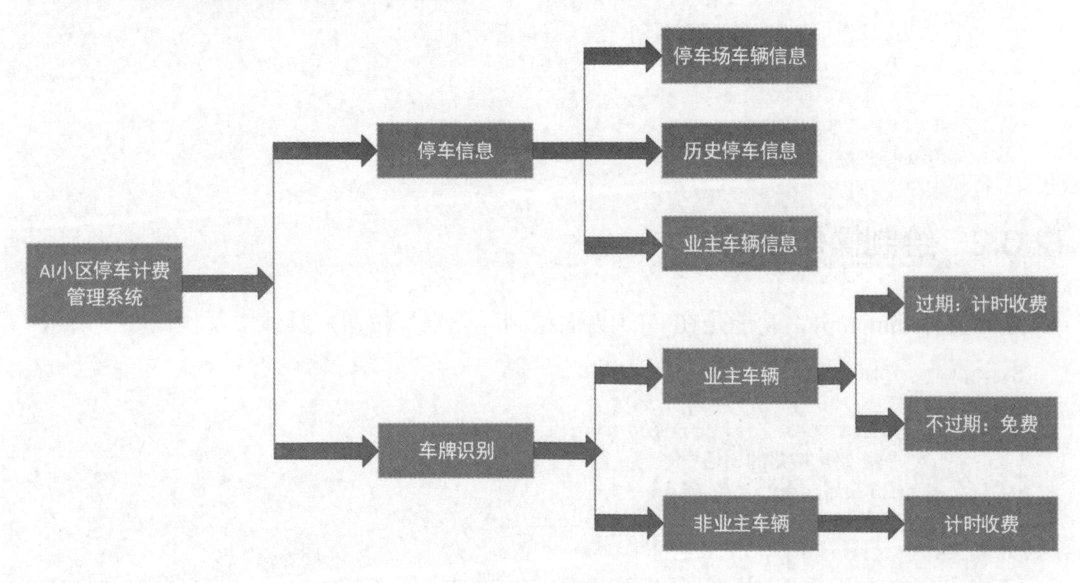

图12-1 功能模块

12.3 系统GUI

本系统是使用库Pygame实现的桌面项目，通过可视化桌面识别进出小区的车辆信息，并自动实现计费功能。

扫码观看本节视频讲解

12.3.1 设置基本信息

编写文件settings.py，功能是设置Pygame桌面的基本信息，包括屏幕大小、颜色和停车位总数等。主要实现代码如下所示：

```
class Settings():
   def __init__(self):
      """ 初始化设置 """
      # 屏幕设置(宽、高、背景色、线颜色)
      self.screen_width = 1000
      self.screen_height = 484
      self.bg_color = (255, 255, 255)

      # 停车位总数
      self.total = 100

      # 识别颜色、车牌号、进来时间、出入场信息
      self.ocr_color = (212, 35, 122)
      self.carnumber = ''
      self.comeInTime = ''
      self.message = ''
```

12.3.2 绘制操作按钮

编写文件 button.py，功能是在 GUI 界面绘制“识别”按钮，具体实现代码如下所示：

```
import pygame.font
class Button():
    def __init__(self, screen, msg):
        """初始化按钮的属性"""
        self.screen = screen
        self.screen_rect = screen.get_rect()
        # 设置按钮的尺寸和其他属性
        self.width, self.height = 100, 50
        self.button_color = (0, 120, 215)
        self.text_color = (255, 255, 255)
        self.font = pygame.font.SysFont('SimHei', 25)
        # 创建按钮的 rect 对象，并使其居中
        self.rect = pygame.Rect(0, 0, self.width, self.height)
        # 创建按钮的 rect 对象，并设置按钮中心位置
        self.rect.centerx = 640 - self.width / 2 + 2
        self.rect.centery = 480 - self.height / 2 + 2
        # 按钮的标签只需创建一次
        self.prep_msg(msg)
    def prep_msg(self, msg):
        """将 msg 渲染为图像，并使其在按钮上居中"""
        self.msg_image = self.font.render(msg, True, self.text_color,
                                  self.button_color)
```

```
        self.msg_image_rect = self.msg_image.get_rect()
        self.msg_image_rect.center = self.rect.center
    def draw_button(self):
        # 绘制一个用颜色填充的按钮，再绘制文本
        self.screen.fill(self.button_color, self.rect)
        self.screen.blit(self.msg_image, self.msg_image_rect)
```

12.3.3 绘制背景和文字

编写文件 textboard.py，功能是绘制 GUI 的背景和文字，分别设置背景图案、绘制线条和识别结果文字的属性。具体实现代码如下所示：

```
import pygame.font

# 线颜色
line_color = (0, 0, 0)
# 显示文字信息时使用的字体设置
text_color = (0, 0, 0)

def draw_bg(screen):
    # 背景文字图案
    bgfont = pygame.font.SysFont('SimHei', 15)
    # 绘制横线
    pygame.draw.aaline(screen, line_color, (662, 30), (980, 30), 1)
    # 渲染为图片
    text_image = bgfont.render('识别信息： ', True, text_color)
    # 获取文字图像位置
    text_rect = text_image.get_rect()
    # 设置文字图像中心点
    text_rect.left = 660
    text_rect.top = 370
    # 绘制内容
    screen.blit(text_image, text_rect)

# 绘制文字(text-文字内容、xpos-x 坐标、ypos-y 坐标、fontSize-字体大小)
def draw_text(screen, text, xpos, ypos, fontsize, tc=text_color):
    # 使用系统字体
    xtfont = pygame.font.SysFont('SimHei', fontsize)
    text_image = xtfont.render(text, True, tc)
    # 获取文字图像位置
    text_rect = text_image.get_rect()
    # 设置文字图像中心点
    text_rect.left = xpos
    text_rect.top = ypos
```

```
    # 绘制内容
    screen.blit(text_image, text_rect)
```

12.4 车牌识别和收费

车牌识别是本系统的核心，为了提高开发效率，本项目使用百度在线 AI 技术识别车牌。

扫码观看本节视频讲解

12.4.1 登记业主的车辆信息

在表格文件“业主车辆信息表.xlsx”中登记业主的车辆信息，包括车牌号和到期时间，如图 12-2 所示。

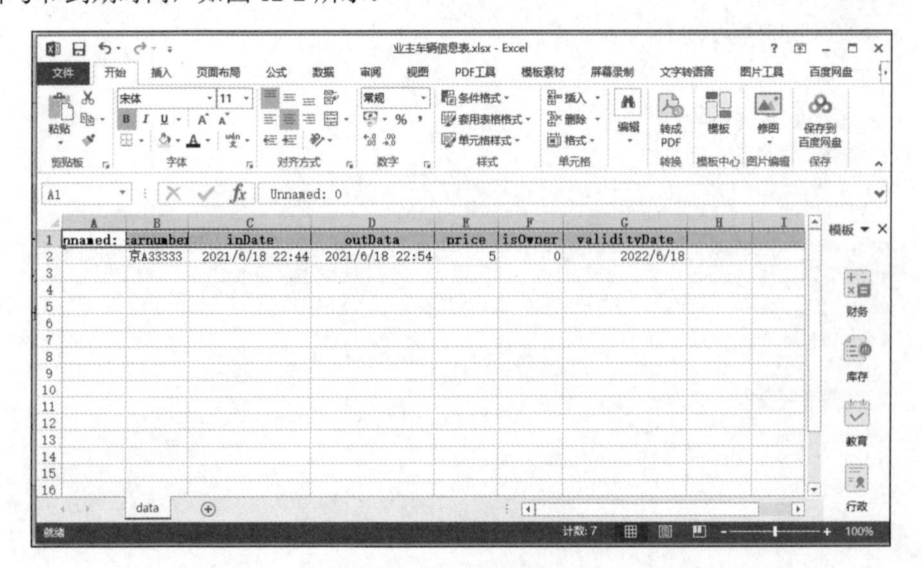

图 12-2 登记业主的车辆信息

12.4.2 识别车牌

编写文件 ocrutil.py，功能是调用百度 AI 中的文字识别 SDK 获取图片中的车牌信息。注意，在测试时图片文件 test.jpg 表示从摄像头读取的图片，每次循环获取一次，大家也可以使用 MP4 视频进行测试。具体实现代码如下所示：

```
from aip import AipOcr
import os

# 百度识别车牌
# 申请地址 https://login.bce.baidu.com/
filename = 'file/key.txt'  # 记录申请的Key的文件位置
if os.path.exists(filename):  # 判断文件是否存在
   with open(filename, "r") as file:  # 打开文件
      dictkey = eval(file.readlines()[0])  # 读取全部内容转换为字典
      # 以下获取的三个Key是进入百度AI开放平台的控制台的应用列表里创建应用得来的
      APP_ID = dictkey['APP_ID']  # 获取申请的APIID
      API_KEY = dictkey['API_KEY']  # 获取申请的APIKEY
      SECRET_KEY = dictkey['SECRET_KEY']  # 获取申请的SECRETKEY
else:
   print("请先在file目录下创建key.txt，并且写入申请的Key！格式如下："
        "\n{'APP_ID':'申请的APIID', 'API_KEY':'申请的APIKEY', 'SECRET_KEY':
                                                    '申请的SECRETKEY'}")
# 初始化AipOcr对象
client = AipOcr(APP_ID, API_KEY, SECRET_KEY)

# 读取文件
def get_file_content(filePath):
   with open(filePath, 'rb') as fp:
      return fp.read()

# 根据图片返回车牌号
def getcn():
   # 读取图片
   image = get_file_content('images/test.jpg')
   # 调用车牌识别
   results = client.licensePlate(image)['words_result']['number']
   # 输出车牌号
   return results
```

12.4.3 计算停车时间

编写文件timeutil.py，功能是根据是否为业主车辆来计算停车时间，主要包含如下两个功能函数。

- 函数time_cmp()：比较出场时间与卡有效期，判断业主是否需要收费。
- 函数priceCalc()：用来计算停车时间，里面存在两种情况，一种是外来车，只需要比较出入场时间差；另一种是业主车，入场时，卡未到期，但出场已经到期，所

以需要比较卡有效期和出场时间的差值。

文件 timeutil.py 的具体实现代码如下所示：

```
import datetime
import math

# 计算两个日期大小
def time_cmp(first_time, second_time):
   # 由于有效期获取后会有小数数据
   firstTime = datetime.datetime.strptime(str(first_time).split('.')[0],
               "%Y-%m-%d %H:%M:%S")
   secondTime = datetime.datetime.strptime(str(second_time), "%Y-%m-%d %H:%M")
   number = 1 if firstTime > secondTime else 0
   return number

# 计算停车时间
def priceCalc(inDate, outDate):
   if '.' in str(inDate):
      inDate = str(inDate).split('.')[0]
      inDate = datetime.datetime.strptime(inDate, "%Y-%m-%d %H:%M:%S")
      print('特殊处理')
   else:
      inDate = datetime.datetime.strptime(inDate, "%Y-%m-%d %H:%M")
   outDate = datetime.datetime.strptime(str(outDate), "%Y-%m-%d %H:%M")
   rtn = outDate - inDate
   # 计算停车多少小时(往上取整)
   y = math.ceil(rtn.total_seconds() / 60 / 60)
   return y
```

注意：在上述代码中，由于读取 Excel 的卡有效期字段，会多出“.xxxx”这部分，所以需要经过 split('.')处理。

12.4.4 识别车牌并计费

编写文件 procedure_functions.py，当单击“识别”按钮后实现识别车牌并计算停车费，主要分为如下两种处理逻辑。

(1) 当停车场未有停车时，只需要识别后，把车辆信息存入“停车场车辆表”并把相关信息显示到界面右下角。

(2) 当停车场已有停车时，会出现两种情况，一种是入场，一种是出场。

在车辆入场时，需判断是否停车场已满，已满则不给进入并显示提示信息，如果未满则需要把车辆信息存入“停车场车辆表”并把相关信息显示到界面右下角。

在车辆出场时，分业主有效、业主过期和外来车 3 种情况收费，删除车辆表相应的车辆信息，并把车辆信息和收费信息等存入“停车场历史表”(可用于后面数据的汇总统计)。

文件 procedure_functions.py 的具体实现代码如下所示：

```
import sys
import pygame
import time
import pandas as pd
import ocrutil
import timeutil

# 事件
def check_events(settings, recognition_button, ownerInfo_table, carInfo_table,
                 history_table, path):
    """ 响应按键和鼠标事件 """
    for event in pygame.event.get():
        if event.type == pygame.QUIT:
            sys.exit()
        elif event.type == pygame.MOUSEBUTTONDOWN:
            mouse_x, mouse_y = pygame.mouse.get_pos()
            button_clicked = recognition_button.rect.collidepoint(mouse_x, mouse_y)
            if button_clicked:
                try:
                    # 获取车牌
                    carnumber = ocrutil.getcn()

                    # 转换当前时间的格式，例如： 2022-12-11 16:18
                    localtime = time.strftime('%Y-%m-%d %H:%M', time.localtime())
                    settings.carnumber = '车牌号码：' + carnumber

                    # 判断进入车辆是否为业主车辆
                    # 获取业主车辆信息(只显示卡未过期)
                    ownerInfo_table = ownerInfo_table[ownerInfo_table
                                      ['validityDate'] > localtime]
                    owner_carnumbers = ownerInfo_table[['carnumber',
                                       'validityDate']].values
                    carnumbers = ownerInfo_table['carnumber'].values
                    # 获取车辆表信息
                    carInfo_carnumbers = carInfo_table[['carnumber', 'inDate',
                                         'isOwner', 'validityDate']].values
                    cars = carInfo_table['carnumber'].values
                    # 增加车辆信息
                    append_carInfo = {
```

```
            'carnumber': carnumber
        }
        # 增加历史信息
        append_history = {
            'carnumber': carnumber
        }
        carInfo_length = len(carInfo_carnumbers)
        # 车辆表未有数据
        if carInfo_length == 0:
            print('目前车辆进入小区')
            in_park(owner_carnumbers, carnumbers, carInfo_table,
                    append_carInfo, carnumber, localtime, settings, path)
        # 车辆表有数据
        else:
            if carnumber in cars:
                # 出停车场
                i = 0
                for carInfo_carnumber in carInfo_carnumbers:
                    if carnumber == carInfo_carnumber[0]:
                        if carInfo_carnumber[2] == 1:
                            if timeutil.time_cmp(carInfo_carnumber[3],
                                                 localtime):
                                print('业主车，自动抬杠')
                                msgMessage = '业主车，可出停车场'
                                parkPrice = '业主卡'
                            else:
                                print('这是业主车，但是已经过期，需要收费抬杠')
                                # 比较卡有效期时间
                                price = timeutil.priceCalc
                                        (carInfo_carnumber[3], localtime)
                                msgMessage = '停车费用：' + str(5 * int(price))
                                             + '(业主您好，您的卡已到期)'
                                parkPrice = 5 * int(price)

                        else:
                            print('外来车，收费抬杠')
                            # 比较入场时间
                            price = timeutil.priceCalc
                                    (carInfo_carnumber[1], localtime)
                            msgMessage = '停车费用：' + str(5 * price)
                            parkPrice = 5 * int(price)

                        print(i)
                        carInfo_table = carInfo_table.drop([i])
                        # 增加数据到历史表
```

```
                        append_history['inDate'] = carInfo_carnumber[1]
                        append_history['outData'] = localtime
                        append_history['price'] = parkPrice
                        append_history['isOwner'] = carInfo_carnumber[2]
                        append_history['validityDate'] =
                                                 carInfo_carnumber[3]
                        history_table = history_table.append
                                 (append_history, ignore_index=True)

                        settings.comeInTime = '出场时间: ' + localtime
                        settings.message = msgMessage

                        # 更新车辆表和历史表
                        pd.DataFrame(carInfo_table).to_excel(path
                            + '停车场车辆表' + '.xlsx', sheet_name='data',
                                                 index=False, header=True)
                        pd.DataFrame(history_table).to_excel(path
                            + '停车场历史表' + '.xlsx', sheet_name='data',
                                                 index=False, header=True)
                        break
                    i += 1
            else:
                # 入停车场
                print('有车辆表数据入场')
                if carInfo_length < settings.total:
                    in_park(owner_carnumbers, carnumbers, carInfo_table,
                            append_carInfo, carnumber,
                         localtime, settings, path)
                else:
                    print('停车场已满')
                    settings.comeInTime = '进场时间: ' + localtime
                    settings.message = '停车场已满, 无法进入'
        except Exception as e:
            print("错误原因: ", e)
            continue
        pass


# 车辆入停车场
def in_park(owner_carnumbers, carnumbers, carInfo_table, append_carInfo,
            carnumber, localtime, settings, path):
    if carnumber in carnumbers:
        for owner_carnumber in owner_carnumbers:
            if carnumber == owner_carnumber[0]:
                print('业主车, 自动抬杠')
```

```
            msgMessage = '提示信息：业主车，可入停车场'
            append_carInfo['isOwner'] = 1
            append_carInfo['validityDate'] = owner_carnumber[1]
            # 退出循环
            break
    else:
        print('外来车，识别抬杠')
        msgMessage = '提示信息：外来车，可入停车场'
        append_carInfo['isOwner'] = 0
    append_carInfo['inDate'] = localtime
    settings.comeInTime = '进场时间：' + localtime
    settings.message = msgMessage
    # 添加信息到车辆表
    carInfo_table = carInfo_table.append(append_carInfo, ignore_index=True)
    # 更新车辆表
    pd.DataFrame(carInfo_table).to_excel(path + '停车场车辆表' + '.xlsx',
                sheet_name='data', index=False, header=True)
```

12.5 主程序

本项目的主程序文件是 main.py，功能是编写主函数初始化程序，调用上面的功能函数展示 GUI 桌面，并监听用户单击按钮实现车辆识别和计费处理。文件 main.py 的具体实现代码如下所示：

扫码观看本节视频讲解

```
import pygame
import cv2
import os
import pandas as pd
# 引入自定义模块
from settings import Settings
from button import Button
import textboard
import procedure_functions as pf

def run_procedure():
    # 获取文件的路径
    cdir = os.getcwd()
    # 文件夹路径
    path = cdir + '/file/'
    # 读取路径
    if not os.path.exists(path + '停车场车辆表' + '.xlsx'):
```

```
    # 车牌号 进入时间 离开时间 价格 是否业主
    carnfile = pd.DataFrame(columns=['carnumber', 'inDate', 'outData',
                            'price', 'isOwner', 'validityDate'])
    # 生成 xlsx 文件
    carnfile.to_excel(path + '停车场车辆表' + '.xlsx', sheet_name='data')
    carnfile.to_excel(path + '停车场历史表' + '.xlsx', sheet_name='data')

settings = Settings()
# 初始化并创建一个屏幕对象
pygame.init()
pygame.display.set_caption('智能小区车牌识别系统')
ic_launcher = pygame.image.load('images/icon_launcher.png')
pygame.display.set_icon(ic_launcher)
screen = pygame.display.set_mode((settings.screen_width,
                                  settings.screen_height))

try:
    # cam = cv2.VideoCapture(0) # 开启摄像头
    cam = cv2.VideoCapture('file/test.mp4')
except:
    print('请连接摄像头')
# 循环帧率设置
clock = pygame.time.Clock()
running = True
# 开始主循环
while running:
    screen.fill(settings.bg_color)
    # 从摄像头读取图片
    sucess, img = cam.read()
    # 保存图片，并退出。
    if sucess:
        cv2.imwrite('images/test.jpg', img)
    else:
        # 识别不到图片或者设备停止，则退出系统
        running = False
    # 加载图像
    image = pygame.image.load('images/test.jpg')
    # 设置图片大小
    image = pygame.transform.scale(image, (640, 480))
    # 绘制视频画面
    screen.blit(image, (2, 2))

    # 创建识别按钮
    recognition_button = Button(screen, '识别')
    recognition_button.draw_button()
```

```
        # 读取文件内容
        ownerInfo_table = pd.read_excel(path + '住户车辆表.xlsx',
                            sheet_name='data')
        carInfo_table = pd.read_excel(path + '停车场车辆表.xlsx',
                            sheet_name='data')
        history_table = pd.read_excel(path + '停车场历史表.xlsx',
                            sheet_name='data')
        inNumber = len(carInfo_table['carnumber'].values)
        # 绘制背景
        textboard.draw_bg(screen)
        # 绘制信息标题
        textboard.draw_text(screen, '共有车位：' + str(settings.total)
                    + '  剩余车位：' + str(settings.total - inNumber), 680, 0, 20)
        # 绘制信息表头
        textboard.draw_text(screen, ' 车牌号          进入时间', 700, 40, 15)
        # 绘制停车场车辆前十条信息
        carInfos = carInfo_table.sort_values(by='inDate', ascending=False)
        i = 0
        for carInfo in carInfos.values:
            if i >= 10:
                break
            i += 1
            textboard.draw_text(screen, str(carInfo[1])+'    '+str(carInfo[2]),
                                700, 40 + i * 30, 15)
        # 绘制识别信息
        textboard.draw_text(screen, settings.carnumber, 660, 400, 15,
                            settings.ocr_color)
        textboard.draw_text(screen, settings.comeInTime, 660, 422, 15,
                            settings.ocr_color)
        textboard.draw_text(screen, settings.message, 660, 442, 15,
                            settings.ocr_color)

        """ 响应鼠标事件 """
        pf.check_events(settings, recognition_button, ownerInfo_table,
                        carInfo_table, history_table, path)
        pygame.display.flip()
        # 控制游戏最大帧率为 60
        clock.tick(60)
    # 关闭摄像头
    cam.release()
run_procedure()
```

执行代码后单击“识别”按钮，可以识别出当前进入小区的车辆信息，如图 12-3 所示。

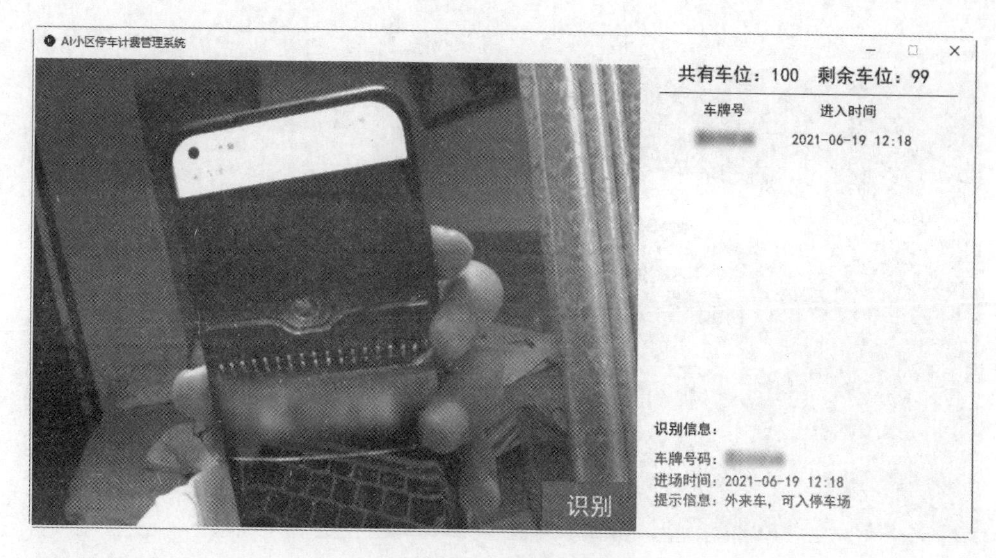

图 12-3 车辆进小区

当车辆出小区时可以实时识别出车辆信息，并计算出停车费用，如图 12-4 所示。

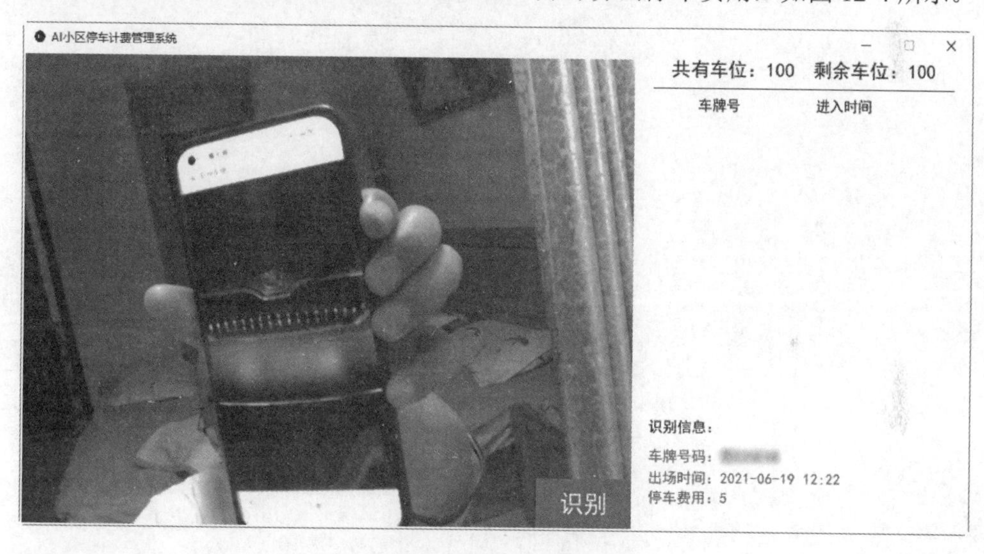

图 12-4 车辆出小区计费

如果是有效期内的业主车辆则免费，如图 12-5 所示。

图 12-5 有效期内的业主车辆